Cambridge astrophysics series

T0296349

Quasar astronomy

Quasar astronomy

DANIEL W. WEEDMAN

The Pennsylvania State University

The right of the
University of Cambridge
to print and sell
all manner of books
was granted by
Henry VIII in 1534.
The University has printed
and published continuously
since 1584.

CAMBRIDGE UNIVERSITY PRESS
Cambridge
New York   New Rochelle
Melbourne   Sydney

CAMBRIDGE UNIVERSITY PRESS
Cambridge, New York, Melbourne, Madrid, Cape Town, Singapore, São Paulo, Delhi

Cambridge University Press
The Edinburgh Building, Cambridge CB2 8RU, UK

Published in the United States of America by Cambridge University Press, New York

www.cambridge.org
Information on this title: www.cambridge.org/9780521303187

First published 1986
First paperback edition 1988
Reprinted 1988

A catalogue record for this publication is available from the British Library

Library of Congress Cataloguing in Publication data
Weedman, Daniel W.
Quasar astronomy.
Bibliography: p.
Includes index.
1. Quasars. 2. Radio astronomy. I. Title.
QB860.W44   1986   523   85-30924

ISBN 978-0-521-30318-7 hardback
ISBN 978-0-521-35674-9 paperback

Transferred to digital printing 2008

To Diana and Sylvia

CONTENTS

PREFACE

The discovery of quasars nearly a quarter of a century ago made a new science out of astronomy. There were two factors in this invigorating revolution. One was the conceptual shock of learning that some very important sources of energy exist in the universe that are not related to the nuclear fusion processes in stars. The other was the fact that the discovery was made with a new technology, in this case radio astronomy. For the theorist, there was suddenly an open season for wide ranging and creative speculations on cosmological processes, energy generation, and radiation physics. For the technologist, there was proof that opening new observational spectral windows could reveal extraordinary and totally unanticipated things. Radio astronomy was quickly followed by ultraviolet, infrared and X-ray astronomy.

That first phase of the theoretical and technical regeneration of astronomy is now complete. Astronomers are more open to heretical theoretical suggestions and unconventional observational techniques. Exceptional telescope facilities are at our disposal worldwide and in space. Two decades of effort have not answered all of the fundamental questions about quasars, nor have they led to the discovery of objects any more puzzling. We still have the problems, but we now have the tools, and so can get on with the work of learning what, where, when, and why are the quasars.

My initiation into the subject began during a few spare hours left over from another project while observing with the 36-inch telescope at McDonald Observatory, in the fall of 1967. I obtained a few exploratory spectra of the Seyfert galaxy NGC 1068 which proved that the analytical techniques of emission line spectroscopy, taught to me so thoroughly by Donald Osterbrock, would have very exciting utility. On the basis of that brief observational experience, Harlan Smith assigned me to accompany

Edward Khachikian to McDonald for the first spectroscopic observations of galaxies found in B. E. Markarian's new survey. As five of the first ten galaxies in Markarian's list proved to be Seyfert galaxies, I was lured into this research area permanently; it was suspected at the time that Seyfert galaxies and quasars had many similarities. (I already owed a debt to the late Carl Seyfert, who had given me my first encouragement in astronomy when I was a high school student in Nashville, Tennessee.) While this research stimulated my own learning about quasars, this book is a product of my attempts to explain them to students in various classes at the University of Texas, Vanderbilt University, University of Minnesota, and Pennsylvania State University.

The primary purpose of this book is to present sufficient equations, definitions of working units, and analysis concepts for a new student of the subject to be able to learn how to transform observations into meaningful conclusions regarding various properties of quasars. Much of the material is useful for applying to observations of galaxies in general, although I present the material only as it relates to quasars. The techniques described include many of those necessary for any research in extragalactic astronomy. Enough recent results are summarized and interpreted for the reader to discern the fundamental conclusions which have now been reached concerning the nature of quasars and their relation to the observed universe. Nevertheless, I attempt to call attention often to outstanding research problems and areas in which many contributions are yet to be made. If readings of this book stimulate a few PhD dissertations, I will feel that a valuable purpose has been served.

I attempted while writing to include sufficient references to guide entry into the extensive research literature. Once into that literature, there are many branching trails one can follow. This book is not a comprehensive review volume for all of those trails. Giving thorough citations to all who work in the field would require a democratic treatment of detailed nuances for many complex issues. I wish to concentrate on first-order conclusions and controversies; some important lessons have been learned about quasar research. I feel it better to summarize the status of our current knowledge and concerns than to distract the reader with all conceivable alternatives.

Many people have had very clever ideas about quasars. Some people have occasionally had silly ideas. All have wrestled with the subject commendably. I encourage the reader to pursue the research literature and become acquainted with the personalities in this research for the past years. Possibly, the interpretations that we now accept contain some very basic flaws, and the present understanding may someday be proven grossly wrong. Either that eventuality or confirmation of the present ideas cannot happen unless competent researchers continue to carry us forward. I hope this book encourages some who will do so.

Daniel W. Weedman
October, 1985
State College, Pennsylvania

1

THE TECHNOLOGY

1.1 Introduction

In the chapters that follow, nearly a quarter century of intensive research and substantial progress in understanding the quasars will be summarized. From the perspective of an astronomer, it would be satisfying to report that this progress was primarily attributable to the cleverness and diligence of astronomers in attacking the problem. To be honest, however, most of the progress should be credited to the engineers and physicists who have developed the tools that allow our wide ranging probes into the mysteries of quasars. Trying to understand quasars forced astronomers into realizing that observations must extend over the broadest possible spectral coverage, that we must learn how to use X-ray, ultraviolet, optical, infrared and radio astronomy. It is now possible for an individual astronomer to have access to telescopes that access all of these spectral regions, and I am convinced that the definition of a 'good' observer in the next few decades will weigh heavily on the ability to be comfortable with all of these techniques. To realize how much this has changed the science of astronomy, one need only recall that the analogously important talents in 1960 were the ability to work efficiently at night, to withstand cold temperatures, and to develop photographic emulsions without accidently turning on the lights. At that time, the technically sophisticated astronomer was one who could use a photomultiplier tube.

In many ways, therefore, the decades of the quasars have also required intensive learning experiences on the part of astronomers. To their credit, many diverted substantial time into making the new tools work and applying them to frontier research problems. It is much harder to do that in the middle of the night than to sit and watch while photons bounce off photographic plates. It is to be assumed that this broadening of skills will continue. Consequently, the discussion in this volume will not only summarize existing results from various techniques, but will refer to the potentialities of new instruments that are only being planned at the time of this writing. The remainder of this chapter describes the quantitative capabilities of those existing and planned telescopes that are most relevant to quasar research. No effort is made to describe all telescopes that have or will contribute, nor is there any credit given to the historical development of how these marvellous devices came to be. What follows is a condensed user's guide to the capabilities of instruments which push the frontiers of quasar research, to help in planning or evaluating observations such as those described in later chapters.

Citations to technical capabilities are absent in this chapter. Partly, this is because such technical descriptions can be tracked down by using references to various observations given in other chapters. It is also partly because such references are often not available as well defined sources. Many of the quantitative values given below just arise from personal experience with various instruments; others are found in technical reports or proposals of limited circulation.

Throughout various discussions, a mixture of units will be used to describe location in the spectrum and the intensity of radiation recorded. This mixing is done without apology. As technologists from other areas applied their talents to astronomy, they brought with them different systems of units. It is not appropriate to attempt an imposition of a standard system on all of these areas. It is a matter of courtesy, much like the attempt to use the local language while travelling. All of the necessary translations are listed in Table 1.1.

Certain definitions and clarifications apply uniformly to all sections below in which spectroscopy is discussed. Describing the achievable flux limit for a spectroscopic observation depends upon the resolution and signal to noise (S/N) of the observation. For quasar observations, there are three categories of resolutions that are relevant in ultraviolet, optical and infrared spectroscopy. If it is only desired to observe the flux level of the continuum and to find the strongest emission lines for determining a redshift, resolution of ~ 100 or worse is adequate. Resolution is defined as the ratio $\lambda/\Delta\lambda$, for λ the observing wavelength and $\Delta\lambda$ the separation of resolvable features. A given value of resolution defines a fixed velocity resolution in the spectrum; resolution of 100 means that features separated by 3000 km s^{-1} are resolved. I define resolution of 100 as 'low' resolution, that of 1000 as 'moderate' resolution, and of 10 000 as 'high resolution'. This is a natural breakdown for quasar astronomy. Most observations are at moderate resolution, because this is adequate to divide emission lines into 'broad' or 'narrow' lines and is adequate to reveal reasonably faint emission or absorption features on the continuum. High resolution is needed only for studying very fine details of emission line profiles, or for studying weak absorption features. The latter are often found with intrinsic line widths of order 10 km s^{-1}, so they can only be described with high resolution observations.

For a spectroscopic observation, S/N achieved is quoted in terms of that per resolution element. This does not have to be high to give a nice looking spectrum, especially if the spectrum can be smoothed through several adjacent resolution elements. Few observations achieve S/N of 10; for relating flux limits below, I target a S/N resolution element of ~5. The S/N achieved increases only with the square root of exposure time, which explains the general preference to achieve many relatively poor observations compared to one good one when observing time is tight.

1.2 Radio astronomy

Beginning with radio astronomy is not meant to indicate that the spectrum will be considered in order of increasing frequency. Radio astronomy comes first here in recognition of the fact that this tool first showed the existence of quasars. Even though it is now clear that the great majority of quasars are invisible to radio telescopes, and even though quasars would somehow surely have been found in the meantime, a great

Table 1.1. *Equivalence of units for spectral and flux measurements*

Unit	is equivalent to	as a measure of
10 cm	3000 MHz	wavelength or frequency
1 Å	10^{-8} cm	wavelength
1 nm	10 Å	wavelength
1 keV	12.4 Å	energy or wavelength
10 keV	1.2 Å	energy or wavelength
$1 \mu\text{m}$	10^4 Å	wavelength
1 Jy	10^3 mJy	flux
1 mJy	$10^{-29} \text{ W m}^{-2} \text{ Hz}^{-1}$	flux
1 mJy	$10^{-26} \text{ erg cm}^{-2} \text{ s}^{-1} \text{ Hz}^{-1}$	flux
$1 \text{ keV cm}^{-2} \text{ s}^{-1} \text{ keV}^{-1}$	$6.6 \times 10^{-27} \text{ erg cm}^{-2} \text{ s}^{-1} \text{ Hz}^{-1}$	flux

Zero magnitude = flux from the star Vega at any wavelength, or $3.4 \times 10^{-9} \text{ erg cm}^{-2} \text{ s}^{-1} \text{ Å}^{-1}$ at 5500 Å, or $3.5 \times 10^{-20} \text{ erg cm}^{-2} \text{ s}^{-1} \text{ Hz}^{-1}$ at $5.4 \times 10^{14} \text{ Hz}$. A factor of 100 less flux corresponds to an increase of five magnitudes.

debt is owed to the radio astronomers for proving what unanticipated results could arise from looking at the universe at a new frequency. The early discoveries of radio astronomy stimulated much of the other technology development in astrophysics. If radio astronomers could find something as exciting and unanticipated as quasars upon taking their first look, what might be seen with other new techniques? While some of the other techniques may prove to contribute more to deciphering quasars, the uses of radio astronomy are hardly outdated.

1.2.1   The very large array: VLA

The VLA is the most sensitive device for determining the fluxes of quasars at wavelengths from 2 to 20 cm. Angular resolution is canonically 1″, although like any interferometer resolution scales inversely with the maximum baseline separation. Design of the VLA is such that the same spatial resolution is achievable at 2, 6 and 20 cm for easy determination of spectra in extended sources. As a result, the largest array configuration (A array) produces the same resolution at 20 cm as does the B array at 6 cm or the C array at 2 cm. In practice, the best feasible resolution for quasars is with the A array at 6 cm, yielding sub-arcsecond resolution. Atmospheric instabilities limit 2 cm usage. The VLA is scheduled in a fixed array configuration for three month intervals, so complete spectral coverage requires more than one session.

Observations require frequent checks of known, standard sources for phase and flux calibration. A typical observation utilizes ten minutes on the source of interest, followed by five minutes on a calibrator, etc., until the total desired integration time is reached. For integration times of order one hour, it is usually possible to detect point sources (spatially unresolved sources) down to 0.1 mJy. The detection limit is more sensitive to the presence of nearby bright, contaminating sources than to the integration time. The weak higher orders of the antenna interference pattern (sidelobes) respond to strong sources well out of the primary beam for which maps are made. If sources cannot be located and subtracted out (the process of 'cleaning', which is the most computer intensive phase of producing a map), residual background noise restricts the detectability of faint sources.

It is feasible to produce maps covering 30′ by 30′ with a single observation, although the limitations of processing hardware limit information content to 1024 by 1024 picture elements (pixels) in most maps. Covering large areas thereby requires sacrificing spatial resolution as a tradeoff against processing time. A more common mode is to map the immediate vicinity of the object of interest using the highest spatial resolution the array allows in a 512 by 512 pixel map. Nevertheless, the wide field 'survey' mode has a good deal of potential for quasar studies in determining radio detections or limits for large numbers of quasars detectable with other techniques.

For a handful of radio bright quasars, especially those with jets and extended lobes, the greatest challenge is to detect weak or low-surface brightness structure close to very bright portions of a source. That is, it is desirable to achieve large dynamic range. This is another sophisticated cleaning problem, which can utilize the nearby bright source to remove sidelobes very precisely. This is the technique of 'self-calibration', onerous only to those whose simpler projects are bogged down in the computer while someone else is self-calibrating.

It is intended for the foregoing comments to leave the impression that the limitations of the VLA are primarily in data analysis hardware and not in the data acquisition capabilities of the array itself. This is true. It means that further improvements in utilization of the VLA, at least at existing wavelengths, will come primarily from improved hardware capabilities for those who handle the data.

1.2.2 Very long baseline interferometry and very long baseline arrays: VLBI and VLBA

While the individual antennas of the VLA are hard-wired together for data assimilation, it is feasible to undertake radio interferometry using disconnected telescopes that phase reference to local clocks. Data are combined after observation. This is the principle of VLBI, which can utilize antennas separated by thousands of kilometers. The ultimate baseline length on the Earth is set by the requirement that a source be simultaneously visible to different antennas. At observing wavelengths comparable to those of the VLA, VLBI achieves spatial resolution measured in milli-arcseconds. Weak sources are detected only if they are unresolved, so a VLBI 'map' is often only a collection of those unresolved sources within a few arcseconds of each other. As sources sometimes disappear by being resolved out as baselines increase, the choice of wavelength or baseline is somewhat driven by source properties.

Sensitivity to faint fluxes of a VLBI experiment can approach that achievable by the VLA, because the total antenna collecting areas are comparable. The most sensitive experiments are those with the greatest frequency bandpass. How large this bandpass can be depends on the capabilities of the data recording tapes and the correlator that subsequently combines the data on these tapes from all stations making observations. Most sensitive experiments are those with a 'Mark III' correlator, but these experiments are much rarer than the 'Mark II' VLBI network observations.

Existing VLBI networks typically utilize, at most, fewer than ten antennas distributed over different continents. Antennas and receivers are of vastly different quality, and are never distributed geographically in the optimum way. Obviously, it is desirable to have as long a baseline in the N–S as in the E–W direction. For any form of interferometry, rotation of the Earth relocates antennas relative to the direction of a source, and the 'aperture synthesis' resulting from this greatly increases the actual baselines utilized. Radio astronomers describe this in context of the 'uv plane'. Achieving maximum source coverage in the uv plane is the goal of any interferometrist. To do this with VLBI, and also to have many optimally designed antennas, it is desirable to construct arrays specifically for long baseline interferometry. This is the plan of the very long baseline arrays, used here in the generic plural because of intentions to build them on more than one continent. At the time of writing, the single project actually called the VLBA plans ten antennas.

It is instructive that a primary design constraint of the VLBA is the weight of data tapes created at a single observing station – restricted for shipping ease to be less than 50 lb a day. The builders of the VLBA had to design high density recorder heads capable of recording ten million bits per square inch, far beyond the capabilities of commercially available products. Correlator design is such that the data can be processed to where map makers can use it in a time no longer than the actual data acquisition. Hardware limitations on the actual analyses of maps suffer some of the same difficulties as with the VLA.

Only a relatively few quasars will be detectable targets with VLBA. These few will hold some basic keys to understanding the radiation mechanisms, however, because VLBI can give spatial resolution a factor of 10^3 better than that achievable with any other feasible technology. As antennas become capable of mm wave VLBI, resolution of 10^{-4} arcseconds is a realistic objective. It is only with the techniques of VLBI that it has been possible to measure directly the size of the radiation-producing region in a quasar. Further improvements in mapping and monitoring these regions are the primary motivation of constructing VLBA.

1.3   X-ray telescopes

In the X-ray spectral region, as well as in the optical, the brightest quasars are usually those visible in the nuclei of nearby galaxies. Most such quasars are labelled Seyfert 1 galaxies; the distinction is strictly morphological, depending only on whether the spatially unresolved quasar is accompanied by a visible galactic envelope. Throughout this volume, I will use the term quasar to include the nuclei of Seyfert 1 galaxies. This usage has not always been conventional, but is a source of confusion only for objects imaged optically. In X-rays, the accompanying galactic envelopes are not seen and the Seyfert 1 nuclei appear as unresolved as any other quasars. This clarification is made because many of the X-ray observations used in this volume as describing the properties of quasars are reported in the literature as observations of Seyfert galaxies or other galactic nuclei.

What is meant by the X-ray region of the spectrum is ambiguous. A few of the brightest quasars have detectable photons out to energies of many MeV. Systematic data on quasars are limited to energies below 100 keV, however, and most such data represent energies below 10 keV. At the low energy extreme, the cutoff of X-rays is taken to be about 0.2 keV. From there to 1000 Å is the realm of 'far ultraviolet'. As a rule, an 'X-ray' detection of a quasar means that it is observed at energies of a few keV.

That quasars are the predominant category of X-ray sources in the sky was demonstrated by the all-sky surveys of the *Ariel V* and *Uhuru* satellites in the late 1970s. Even while data from those pioneering telescopes were being digested, X-ray astronomy took its next big leap with the flight of the High Energy Astrophysical Observatories, HEAO 1 and HEAO 2. From these two satellites comes most of the quasar data which follow herein.

1.3.1 HEAO 1

The major task of this telescope was to carry out more sensitive all-sky surveys than available from *Ariel V* or *Uhuru*, and to extend these surveys over broader energy ranges. Results yield systematic X-ray data for the brightest quasars in the 2–40 keV bandpass of the HEAO 1 A2 experiment and the 15–100 keV bandpass of the A4 experiment. The former has been most thoroughly analyzed. Broad band detections such as those of HEAO 1 A2 can only be reported in terms of the total flux received through the entire bandpass. Even the limits of the bandpasses are vaguely defined, because the detection is the product of the detector sensitivity folded into the spectrum of the source. Quasar spectra are power laws in the X-ray of approximate shape such that flux per unit frequency interval goes as frequency to the power -0.5 to -1. The incidence of photons (which is what the detectors actually register) goes, therefore, as photon number to the power -1.5 to -2. The net result is that many more lower energy photons are recorded from a quasar in a wide bandpass than are higher energy photons, so the lower ranges of the bandpass are responsible for most of the flux recorded in the detection. The results of the A2 survey are quoted as fluxes in the 2–10 keV bandpass, in which the limiting sensitivity is 3×10^{-11} erg cm^{-2} s^{-1}.

Limited spectral information can be deduced from broad band detections, especially if a power law form is assumed. Detectors can register the number of incoming photons as a function of their energy, and various power laws can be folded through the bandpass to determine the best fit. Comparing detections from other bandpasses increases the spectral information. Most data on quasar spectral forms at energies above 5 keV arise from the HEAO 1 detectors.

1.3.2   HEAO 2: Einstein Observatory

Unlike all previous X-ray satellites, which had spatial resolution of only a few degrees and detected sources by pointing in their general direction, HEAO 2 carried an actual telescope which produced X-ray images. Data are available from two imaging detectors, the high resolution imager (HRI) and the imaging proportional counter (IPC). The differences are that the HRI has much better spatial resolution, being able to locate a source to within a few arcseconds, while the IPC has error circles of order one arcminute. Both are 'soft X-ray' detectors, having bandpasses quoted as 0.3–3.5 keV. Little spectral information is available, but the imaging detectors are capable of long integration times and so can register much fainter quasars than have ever been seen before. Most observations were with the IPC for which exposure times of ~3000 s could reach sources of 10^{-13} erg cm^{-2} s^{-1} over the entire bandpass. Longest exposures, in a few 'deep-survey' fields, were pushed to 50 000 s, going to flux limits of 10^{-14} erg cm^{-2} s^{-1}.

It should be realized that the great majority of X-ray observations of quasars have come from the IPC, so all that is available is a broad band flux. The standard description of X-ray flux is normalized to 2 keV, matching this bandpass.

In addition to the imaging detectors, HEAO 2 carried a non-imaging, non-dispersive spectrometer. This was a solid state spectrometer, the SSS, that was capable of measuring low energy spectra of objects comparable in flux with those seen at higher energies by HEAO 1. The SSS sensitivity extended from 0.5 to 4.5 keV, able to reach limiting flux of about 2×10^{-12} erg cm^{-2} s^{-1} over the bandpass, or 8×10^{-30} erg cm^{-2} s^{-1} Hz^{-1} at 3 keV. Although only available for a limited number of galactic nuclei, the SSS spectra are the highest quality so far available in the X-ray. They are very useful for showing absorption effects at the lower energies.

1.3.3 Advanced X-ray astrophysics facility: AXAF

Although the Einstein imaging detectors were sufficiently sensitive to detect quasars of the 21st optical magnitude, X-ray spectra have not been obtained for quasars fainter than optical magnitude about 16. This means that nothing is known about the X-ray spectra of quasars at high redshifts and high luminosities. Even the direct imaging did not produce counts of X-ray sources to sufficiently faint limits to determine the fraction of the diffuse X-ray background which is attributable to identified sources. Because of the importance of X-ray spectra to understanding the energy generating mechanisms in quasars and to understanding the composition of the background, the foremost objective of X-ray astronomy is a capability for obtaining spectra for faint sources. There are various approaches to achieve this, the most desirable of which would be to provide a permanent X-ray observatory to which all astronomers have routine access. At the time of writing, the instrument which seems most likely to fulfil this objective is AXAF.

Throughout the chapters that follow, experiments are mentioned which can only be done with the capabilities planned for AXAF. The technology being incorporated includes a telescope with more collecting area and better imaging than Einstein, along with detectors with greater efficiency and improved spectral resolution. Because of limitations on imaging telescopes, the X-ray bandpass will not achieve energies as high as HEAO 1 did, but will be an improvement over Einstein. The greatest promise is the intent to use solid state imaging detectors, available in the form of charge coupled devices (CCDs). At X-ray energies, these not only register the energy of each incoming photon, but also locate it in the telescope focal plane.

The result is a wide field image of the X-ray sky that also yields a spectrum of everything in the image. With feasible CCDs, it appears that an AXAF imaging spectrometer can reach a flux limit of $5 \times 10^{-34}\,\mathrm{erg\,cm^{-2}\,s^{-1}\,Hz^{-1}}$ at energies of a few keV. This would produce spectra for objects less than 10^{-4} as bright as those seen by the SSS, or produce spectra for objects less than 10% as bright as the faintest images detected in the Einstein deep imaging surveys. Making an analogy in terms of optical astronomy, this would mean an improvement in spectroscopic capabilities equivalent to going from a 10 cm to a 10 m telescope. It is no surprise, therefore, that any discussion of quasars defers to the promises of AXAF in dealing with fundamental questions about X-ray properties.

1.4 Ultraviolet astronomy

Being able to observe the ultraviolet spectrum of nearby objects is particularly critical to understanding galaxies and quasars. This is because such objects, when observed at high redshifts with ground based optical telescopes, are seen only via their intrinsic ultraviolet spectrum. In order to relate objects of low and high redshift, therefore, it is necessary to observe the low redshift objects in ultraviolet wavelengths. The first space telescopes, the Orbiting Astronomical Observatories (OAO series), were ultraviolet telescopes. Existing and planned instruments maintain a substantial capability for obtaining ultraviolet spectra of quasars.

1.4.1   *International ultraviolet explorer: IUE*

The IUE is only a 40 cm telescope, but it provided yeoman service for a decade and introduced many astronomers to the use of an orbiting telescope. In terms of length of service and variety of astronomical applications, the IUE provided a distinctive landmark in space astronomy. Even when further observations can no longer be made, much useful information remains unexploited in the massive and organized IUE data base.

Equipped only for spectroscopy, the IUE can obtain spectra from 1200 to 3300 Å. The short wavelength limit is set by the contamination of geocoronal Ly α emission, and the long limit matches to what can be reached from the ground. Observations of quasars require long integration times, in a few cases pushed to 16 h. For a benchmark exposure time of 1 h, a flux limit of 5×10^{-14} erg cm^{-2} s^{-1} Å^{-1} is reachable at 2000 Å, with low to moderate resolution. For typical quasar spectra, this corresponds to optical magnitude 14, so it is clear that the IUE is not capable of observing the majority of known quasars. Most of the relevant IUE data discussed herein apply to low redshift quasars which are the nuclei of Seyfert 1 galaxies; in particular, a few systematic monitoring observations of changes in such spectra over time scales from days to years have been an outstanding contribution of the IUE.

1.4.2   *Hubble space telescope: HST*

It is obviously desirable to extend ultraviolet spectroscopic capabilities to much fainter objects than could be observed by IUE. Within the ultraviolet spectral region, this is the major gain of the HST. With 30 times the collecting area, improved imaging qualities and improved detectors, HST can reach eight or more magnitudes fainter on objects with comparable resolution to that of IUE. For quasars, this is faint enough to probe to the highest known redshifts. Low and moderate resolution spectroscopy are achieved with the faint object spectrograph (FOS). At 2000 Å with low dispersion, this can reach 5×10^{-18} erg cm^{-2} s^{-1} Å^{-1}. At moderate resolution, the limit is more like 5×10^{-17} erg cm^{-2} s^{-1} Å^{-1}.

Utilizing the brighter quasars (even some with redshifts above 2.5 are sufficiently bright), the high resolution spectrograph (HRS) can probe intergalactic absorption features all the way from the local universe back to the epoch of the quasar observed. Observing at high resolution, a limit of 2×10^{-15} erg cm^{-2} s^{-1} Å^{-1} can be expected at 2000 Å. Here, as in all other cases reported above, the limit given is not an unbreakable one. Remember that values are quoted for exposures of 1 h, but there is no fundamental limit to increasing total exposure time indefinitely to reach fainter objects, even if it means adding together a number of successive exposures.

1.5   Optical telescopes

Most quasars now known have been discovered with and are visible only to optical telescopes. What this means is that in terms of overall sensitivity, the match between quasar spectra and detection capabilities is optimized for large, ground based optical telescopes compared with that obtained by any other observing technique. In discussing the capabilities of optical telescopes, it is necessary to distinguish between the limits attainable in broad band imagery compared with the limits to which spectroscopy can be done. The former is limited by the background brightness of the night sky; the latter is usually limited by noise in the detector itself. For imagery, therefore, collecting sufficient photons from the source is not a problem with existing telescopes; the problem is accurate subtraction of the background. This accounts for the great gains expected with HST. For moderate to high resolution spectroscopy, more photons per resolution element are needed, which is the primary motivation behind plans to build gigantic ground based telescopes.

1.5.1   *Schmidt telescopes*

A classical technique of astronomy, in use for more than half a century, is to use a telescope with a large field of view to image large areas of the sky on large photographic plates. Surveys of this nature, done with Schmidt telescopes, have provided the fundamental image data base of the night sky. Such telescopes have also discovered most of the known quasars using imagery in different wavebands and searching for objects with characteristic quasar colors. Alternatively, quasar searches have been made using imagery obtained through objective prisms on these telescopes, so the visible spectrum of every imaged object is seen with sufficient detail to pick out quasars.

The bulk of quasar survey work, with either direct imagery or objective prism techniques, has been done with one of five telescopes distributed in both northern and southern hemispheres. Ranging in size from 90 cm to 1.2 m, these telescopes are not large by optical standards, but they have been crucial in accumulating quasar statistics. The detector technology remains as photographic plates. Covering areas of the sky from 25 to 35 square degrees in one exposure, with typical exposure times of one to two hours, the data rate is immense. Considering that the plates used have resolution capability (pixel size) of 10 μm, and areas of order 10^3 cm^2, each plate contains 10^9 pixels of information. The challenge to the observer is in extracting the information of interest; for quasars, this relates to only a small fraction of the objects imaged on the plate.

By utilizing the fine grain Kodak IIIaJ emulsion (sensitive from the atmospheric ultraviolet limit to 5200 Å), together with optimum hypersensitizing techniques, imagery to magnitude 23 is feasible with the largest Schmidt telescopes and conditions of best seeing. (Hypersensitizing uses baking in various gases, usually nitrogen or hydrogen mixtures, to remove all moisture from the emulsion. This is a process carried out within a few days prior to using the plate, and is one of the tricks of black magic that makes old fashioned astronomy seem very qualitative to the uninitiated.) The redward wavelength limit is the greatest single handicap of photography. While the IIIaF emulsion is available to extend to 6800 Å, the wavelength response of this emulsion is more irregular than IIIaJ and thereby harder to calibrate. This is particularly serious in objective prism surveys, where it is important to attribute spectral features to the real nature of the spectrum and not to the emulsion response.

The magnitude limit for objective prism surveys is brighter than for direct imaging because the image is spread into a spectrum imposed on an undiminished background. The amount of this dilution and the consequent sensitivity loss depend on the dispersion used. Most quasar surveys done with objective prisms disperse the spectrum to 1000–2000 Å mm^{-1}. The resolution depends on the image size of a point source; at plate scales of 70″ to 100″ mm^{-1} and seeing disks of 2″, resolution is approximately 0.02 mm, or 20–40 Å. This means that the light which would be in a single broad band image is spread out by a factor of 25 to 50, causing a loss of three to four magnitudes in the faintest image which can be seen in contrast to the background. As a result, the best Schmidt telescopes do not obtain objective prism spectra fainter than 20 magnitude (20 mag). Resolution is adequate to pick out the stronger emission and absorption lines in quasar spectra. In principle, hundreds of quasars should be detectable with such a technique on the best Schmidt plates.

1.5.2   *4 m-class telescopes*

4 m-class telescope will be used here as a generic description of the world's large optical telescopes, ranging in size from 3 to 6 m; less than ten such telescopes throughout the world have contributed significantly to observing quasars. Most such observations have been spectroscopic, for which maximum photon collection is essential. A few of these telescopes are capable of imaging over fields as large as one degree. Imaging on photographic plates under the best seeing conditions can reach 25 mag. Analogous techniques to those used with Schmidt telescopes have been used to undertake quasar surveys utilizing broad band imagery or low resolution spectroscopy. For the latter technique, it is not feasible, of course, to place an objective prism over the entire telescope aperture, so prisms or transmission gratings are placed immediately in front of the focal plane. A dispersive element of this nature is labelled 'grism' or 'grens' (for grating + prism or grating + lens). This provides the faintest survey technique used extensively for quasars, with grens and grism plates capable of detecting quasars via the strong emission lines to 22 mag.

While such spectroscopic surveys can discover fainter quasars with these telescopes than with Schmidt telescopes, it should be noted that resolution, for the same dispersion, is not nearly so good. The reason is the greatly increased image size on plates from the larger telescopes. At the prime foci, such telescopes have plate scales of $10''$–$20''$ mm^{-1}; typically a factor of five more magnification than on a Schmidt plate. For the same size image due to seeing and the same spectral dispersion, this means that a resolution element is five times larger, so a dispersion yielding 50 Å resolution on a Schmidt plate would yield 250 Å resolution with a 4 m-class telescope. The most important lesson of such comparisons is that efficiency of a large telescope for any imaging project is greatly dependent upon seeing conditions; far more is accomplished from a few plates with very good seeing than from a large collection of plates with mediocre seeing. My own experience with grism and grens surveys yields quasar detections per plate above 20 for sub-arcsecond seeing, dropping to zero for $3''$ seeing.

The greatest source of progress in both optical imagery and spectroscopy has come from improvements in detectors rather than in the availability of large telescopes. There are several advantages to electronic detectors, particularly the CCDs, beyond the obvious advantage of producing digital data for each pixel that can be massaged to the ultimate ability of one's image processing hardware. The single greatest advantage of CCD imaging compared with a photograph is linearity over a large dynamic range: 10^3–10^4.

This makes it possible to subtract accurately a very weak signal imposed on a strong background. For this reason, CCD imaging has reportedly detected objects to 27 mag under the best ground based conditions with long integration times and addition of successive exposures. Furthermore, a CCD has smooth wavelength response extending to 8000 Å, so it can be used for red sensitive imagery far more easily than photographic plates. The major disadvantage for imagery is the small size of CCD chips; typical devices rarely have more than 25×10^4 pixels, with pixel size ~ 25 μm.

For spectroscopy, the small chip size is not a major disadvantage as long as spectrographs are designed to match the format of the chip. Using optimally designed spectrographs with the best chips, it is possible to obtain low to moderate resolution spectra to 21 mag with the 4 m-class telescopes in exposure times less than a few hours. Because the quantum efficiency of detectors cannot be much more improved, this is about the feasible limit for spectroscopy with existing ground based telescopes. The limit for higher resolution spectroscopy scales appropriately. It is in the hope of going well beyond this limit that a new generation of very large ground based telescopes is under design.

1.5.3   Next generation telescopes: NGTs

With the completion of a 6 m telescope, it was obvious that the classical mirror-making technology in use throughout the century had reached its limits. To build bigger telescopes required fundamentally new designs. There are two alternative approaches to these designs. One is to concentrate on improvements in mirror casting, making thinner blanks with less glass so as to decrease thermal inertia. (The image quality of a large mirror is fundamentally controlled by temperature differentials within it and between it and the ambient air.) It appears feasible to make single blanks as large as 7 m, although polishing and coating of such blanks is also a challenge. The other approach is to use many small mirrors to fabricate the equivalent of a single large one, using computer controlled sensors to maintain alignment of the individual segments. There are various breeds of multiple mirror telescopes arising from this concept, but the largest design realistically considered is for a 15 m telescope. Assuming such a telescope was available, and scaling from existing experience with 4 m-class instruments, the increase in collecting area by a factor of ten could push the spectroscopic magnitude limit to 25.

It is important to realize that the gains advertised for the class of NGTs require siting these telescopes at locations where seeing is superb. Image scales, of necessity, are so large with these gigantic telescopes that efficient matching to the pixel size of feasible detectors places great demands on input image quality. Accompanying the advances in telescope design must also be advances in understanding and controlling the causes of poor seeing. For ground based astronomers, the atmosphere provides the fundamental limitation to what can be achieved.

1.5.4   Hubble space telescope

It was primarily in hope of overcoming atmospheric limitations in optical imagery that HST was conceived. By utilizing the full diffraction limit of the telescope, without atmospheric image distortions, immense gains can be made in the obtainable magnitude limit. HST carries two imaging systems, either of which should be able to image quasars to 28 mag in exposure times of about one hour; this is the faintest limit mentioned anywhere in the present volume. Furthermore, the angular resolution should reach 0.1″ to 0.2″ in search of galactic envelopes around quasars. This, too, is essential progress in attempting to define the nature of galaxies associated with quasars at high redshift.

The wide field/planetary camera (WFPC) carries four CCDs, each with 800 × 800 pixels. Pixel size is 0.1″, and total area imaged is 2.6′ × 2.6′. A particularly useful aspect of the WFPC is that it can function simultaneously with other instruments. This means that survey projects not requiring particular target directions can proceed while other, targeted objects are being observed in another mode. For quasars, a primary objective of such surveys will be low resolution spectroscopic surveys. The WFPC has a grism yielding resolution comparable with that from Schmidt or 4 m-class surveys mentioned above, but which should reach 25 mag because of the much smaller image size.

The other imaging device is the faint object camera (FOC). This utilizes a highly magnified image detected with a greatly intensified tv tube (i.e. a detector read out with a scanning beam). For quasar research, it is probable that the greatest utility of the FOC will be in high resolution imaging of galaxies around quasars. The field is too small to expect much from a survey mode; the pixel size at smallest magnification is 0.04″ × 0.04″, for a field of view of only 22″ × 22″. The advantage of this format is that spatial resolution will not be at all limited by detector resolution, as will be the case to some degree with the WFPC.

Even though of substantially smaller aperture (2.4 m) than many existing or planned ground based telescopes, HST will be the preferred instrument for faint object spectroscopy in the case of observations that are background limited. This means that for low resolution observations, in which detector noise is insignificant and enough photons are available per resolution element, HST can outperform NGTs. The FOS was designed to obtain optical spectroscopy (at 5000 Å) to 26 mag in the lowest resolution mode. At moderate resolution, however, this limit drops closer to 22 mag (as usual, all numbers for exposures of 1 h). This illustrates the regime in which the great photon collecting ability of the NGTs can obtain spectra that cannot be obtained any other way, especially when such spectra are desired of large numbers of objects.

1.6   Infrared telescopes

Infrared astronomy is more dependent on the quality of detectors than any other observational area. That is, the quality varies widely, and comparable telescopes may not compare at all in infrared detectability because of gross differences in detector quality. It is particularly frustrating when one knows that good detectors exist but are unavailable for distribution. There are many possible ways of doping or otherwise mistreating solid state materials to coax them into producing a detectable electronic signal as a consequence of impinging photons. Many of these ways are yet to be fully explored, so major breakthroughs – especially in producing infrared arrays analogous in operation to CCDs – can be expected.

1.6.1   Ground based infrared telescopes

Assuming a decent detector is available, the only other rule of thumb for infrared astronomy is to keep everything in sight as cold as possible. The fundamental problem in trying to observe infrared sources is, of course, that the instrument, telescope and atmosphere radiate brightly at infrared wavelengths, especially the 'thermal infrared' near 10 μm. It is this background emission that limits ground based infrared astronomy and means that telescope size may not be the primary determinant of success. For detecting quasars or other unresolved sources, however, the aperture of the telescope is important in setting the spatial resolution and, thereby, the contrast between source and background. At wavelengths beyond ~10 μm, atmospheric seeing normally yields smaller images than does the telescope diffraction limit, which is a very different situation than for optical astronomy. This explains much of the great desire on the part of infrared astronomers for NGTs.

Any observation at a wavelength longer than 1 μm but shorter than 1 mm could be called infrared. Sometimes, wavelengths in this range but beyond a few hundred μm are called 'submillimeter'. From the ground, water vapor in the atmosphere prevents observations beyond about 30 μm. A good wavelength to refer calibrations to is 10 μm. At this wavelength, a ground based, broad band observation can reach a flux limit of 100 mJy with integration time of order one hour. (Broad band means that 'resolution' as defined in Section 1.1 is ~10.) This limit is more or less proportional to wavelength between 1 and 30 μm, fainter limits being achievable at the shorter wavelengths. Spectroscopic observations, of which there are quite a lot at the shorter wavelengths for emission lines, would have limits scaled with the resolution achieved.

1.6.2 Infrared astronomical satellite: IRAS

To achieve substantial gains over the infrared capabilities of ground based telescopes, observations from above the atmosphere are necessary. The first great success in achieving these routinely for quasars and related galactic nuclei was with the IRAS. Carrying aloft liquid helium to cool itself (the duration of which determined the mission lifetime), IRAS mapped the sky in four infrared bands: 12, 25, 60 and 100 μm. The published all-sky survey contains 250 000 sources, of which less than one thousand relate to quasars. Limiting sensitivities for the survey in the four bands are, respectively, 0.25, 0.25, 0.5 and 1 Jy. Reprocessing satellite data beyond what was in the published survey can push flux limits at a given location a factor of two or three fainter.

The IRAS data base provides excellent guidelines to the infrared continuum properties of some bright quasars and many related galactic nuclei. These data will help determine which nuclei contain quasars obscured by dust, although the broad band infrared data alone cannot unambiguously locate a quasar without substantial help from optical spectroscopic observations. As is the case with many survey techniques, it is very desirable to have self-contained spectroscopic capabilities. That is, if infrared sources exist which are invisible to any other technique, it will be necessary to analyze these sources only with infrared spectroscopy. To provide this capability is a major goal of the next round of orbiting infrared telescopes. Much fainter detection limits can be reached than for IRAS, because IRAS was not able to integrate on a given source position. (This limitation also explains why ground based telescopes can reach the IRAS limit at the shorter wavelengths.)

1.6.3 Space infrared telescope facility: SIRTF

Design plans for SIRTF represent feasible capabilities that can be achieved with the best helium cooled detector arrays on an orbiting, 1 m-class telescope. Such an infrared observatory is a realistic project although, at the time of writing, launch dates and operational details are undetermined. Nevertheless, for planning long term quasar studies, it is useful to have a benchmark for what SIRTF could do. Its greatest astrophysical utility will be the ability to study faint infrared sources with spectroscopy having various resolutions. In the infrared, all spectroscopy from the ground is background limited, which explains why gains are so dramatic in going aloft.

At some short wavelength, around 3 μm, tradeoffs between a smaller orbiting telescope and an NGT on the ground balance for spectroscopic uses. Plans for SIRTF spectrometry are therefore restricted to the range from about 4 to 200 μm. The latter limit is set in part by detectors and in part by the increasing diffraction size of the telescope mirror. At low resolution, even an orbiting infrared telescope will be background limited with SIRTF-quality detectors. In this case, the background is that of the zodiacal dust. It is expected that a faint flux limit of 0.3 mJy is feasible at 10 μm, rising to only 1 mJy by 100 μm, for integration times of less than an hour. At moderate resolution, the detections become more limited by detector noise, in which case increasing integration times would improve the detection limit. Restricting to times less than one hour, moderate resolution limits are about 5 mJy from 10 to 100 μm.

1.7   Conclusions

When the array of instruments described above are finally available for routine astronomical use, we can expect no more technological breakthroughs that will provide order of magnitude improvements. It was such large advances in one step that provided the motivations for going from single dish radio telescopes to the VLA, from VLBI to VLBA, from IUE to HST, from 4 m telescopes to NGTs, from HEAO to AXAF, and from IRAS to SIRTF. In each case, very large gains could be visualized and realized by taking only the one step to the next instrument. That situation is not likely to occur again in the foreseeable future of astronomy, because the generation of instruments now under design comes close to representing the ultimate feasible limit with realistic technology. It is immensely rewarding and commendable that humankind is willing to commit such resources to understanding its Universe. Future generations of astronomers and all who philosophize about the heavens will owe a great deal to the technologists who pushed our science in the last quarter of the 20th century.

2

QUASAR SURVEYS

2.1   Optical surveys for quasars

Quasars cannot be studied until they are found. The purpose of any quasar survey is simply to provide an efficient method of discovering quasars. This efficiency is greatly enhanced if many quasars can be found with a single observation by the detecting instrument, so it is preferable if the observation has a wide enough field of view to include many detectable quasars. Furthermore, it is desirable but usually not feasible to identify a quasar with the survey observation alone, without the necessity of a subsequent observation with another instrument. Because of their characteristic signatures in many different parts of the spectrum, quasars can be surveyed for using various techniques. Much of the subsequent research effort goes into comparison of results from various techniques, to determine whether the same quasars are being found in different ways, or whether there are categories of quasars conspicuous to one form of observation but invisible to another.

2.1.1   Color based surveys

Quasars are easy to find with optical telescopes; a summary compilation by Smith (1984) lists over 40 surveys. The reason is because their spectral characteristics are so different from most stars and galaxies that broad band optical surveys using only three effective wavelengths can differentiate most quasars from other objects. Despite many other techniques for quasar surveys, including X-ray and radio surveys, the great majority of quasars have been discovered optically, and it is very probable that this will continue to be so. *Ex post facto* observations of quasars found any other way show that, as a rule, such quasars have the same optical characteristics as quasars discovered optically, so it is just as complete to start with the optical surveys. Furthermore, the converse is not true.

Virtually all quasars discoverable optically are invisible to the largest radio telescopes and too faint as well to expect X-ray detections with any X-ray telescope other than AXAF. Infrared surveys of the extragalactic sky are so dominated by dusty galaxies, which can mimic the infrared spectral properties of quasars, that no infrared source can be definitively considered a quasar unless optical follow up observations are made. The optical techniques are so extraordinarily efficient that they will remain the unchallenged source of quasars, at least in terms of number found.

For almost 20 years, the optical hunt for quasars proceeded by examining photographs in search of 'blue stellar objects' (BSOs) or 'ultraviolet excess objects'. These are the same things. Consider a typical quasar continuous spectrum, in the optical a power law of index -0.5 to -1 (Chapter 7). In the classical *UBV* system of measuring stellar colors, the transformations of fluxes to magnitudes (Hayes & Latham 1975) are given by

$$U = -2.5 \log f_{3600} - 21.29,$$
$$B = -2.5 \log f_{4400} - 20.42,$$
$$V = -2.5 \log f_{5550} - 21.17,$$

with fluxes as defined in Chapter 3. When power law continua $f_\lambda \propto \lambda^{-(2+\alpha)}$ are used, the color indices $(U - B)$, $(B - V)$ fall in the range -1.1, $+0.5$ to -1.2, $+0.3$ for $-1 < \alpha < -0.5$, typical of quasar continua. These are very different from the colors of most stars. The definition of magnitude units is based on the star Vega, which is taken to have $U - B = B - V = 0$, with cooler stars having more positive color indices. The distinction of quasars, therefore, is that they have too negative $U - B$ compared with most stars, or an 'ultraviolet excess'. It happens that the stars which mimic quasar colors are mainly hot white dwarfs. A lot of these show up in quasar surveys based only upon colors, but the percentage of contaminating stars drops rapidly with magnitude, exceeding 95% at 16 mag (Schmidt & Green 1983) but under 5% at 22 mag (Koo 1984). It is never possible on the basis of *UBV* colors alone to be positive that a candidate is a quasar without an actual spectrum. The detection statistics improve significantly when a redder color is also incorporated, because the quasar power law continues into the red whereas hot star spectra are dropping rapidly.

The most common stellar contaminants in quasar surveys are white dwarf stars. Now, these stars are looked upon as nuisances serendipitously discovered when the intent was to find quasars. At one time, the status was reversed. The first 'quasar' whose spectrum was ever deliberately obtained was observed with the intent of studying white dwarfs. This quasar, Tonantzintla 202, had been found as a 'star' in 1957 and observed spectroscopically in 1960, a few years before 3C 273 or 3C 48, but interpreted to be an old nova or peculiar white dwarf. It was not suspected as a quasar until 1966 (Greenstein & Oke 1970). One wonders if we are now interpreting some white dwarfs as peculiar quasars.

Surveys based only on color have a serious and worrisome bias as a function of redshift. This bias arises because of strong emission lines affecting colors and distorting them from what they would be from the continuum alone. By far the most serious effects come from the Ly α line, which comes into the optical bands at $z \sim 2$ (then having an observed wavelength of 3600 Å). At such redshifts, Ly α typically has an equivalent width of several hundred angstroms, comparable to that of filter bandpasses, so the flux from it can actually exceed that from the continuum within a single filter. As long as Ly α is within the U band, this just makes the quasar even bluer so does not affect detection on a BSO criterion, but as soon as Ly α creeps into the B band it enhances the B and thereby reddens the $U - B$ color. Such an effect begins to happen at $z \sim 2.1$, so BSO surveys may become very inefficient in finding quasars with higher redshifts. These surveys should give trustworthy counts of all quasars at lower redshifts. But the distribution of quasars as a function of magnitude is more influenced at faint magnitudes by higher redshift quasars, so the magnitude distribution of all quasars will not be accurately depicted to faint magnitudes by BSO surveys alone.

The point is demonstrated in Section 2.4, where the anticipated redshift distribution of quasars to various magnitude limits is shown, based upon the evolutionary model described in Chapter 6. Note that 20% of the quasars expected to have $z < 5$ have $z > 2$ for limit 19, but 54% for limit 22. Color based surveys therefore become progressively more incomplete at fainter magnitudes.

2.1.2 *Spectroscopic surveys*

In part to avoid redshift bias introduced by emission lines, but primarily to find quasars based on the detection of emission lines, numerous low dispersion spectral surveys have been undertaken. Using objective prisms or transmission gratings, such surveys yield spectra at 1000–2000 Å mm^{-1} over the entire field of view of the telescope used (Hoag & Smith 1977, Smith 1981). This is the easiest way to find quasars because the strong emission lines are conspicuous, leaving no doubt that a candidate is a quasar rather than a hot star. But all quasars may not be at such a redshift to have a strong line in the visible range; in that case, confirmation spectra are still needed, but detection of the quasar based on its continuum properties remains just as valid as in a multi-color survey. So there is no doubt that spectral surveys can give more complete samples than color surveys. The disadvantage is that spectral surveys cannot detect as faint an object and are a good deal more trouble to analyze quantitatively. A rule of thumb is that a low dispersion spectrum shows a continuum to a limit about two magnitudes brighter than a broad band filter image with the same telescope. The reason is obvious; the spectral resolution is of order 100 Å, and filters have widths of 500 to 1000 Å, so five to ten times as much flux is concentrated on the same part of the photograph in a broad band image. Spectral plates are also more difficult to search in an automated fashion, a much more intelligent computer being required to analyze an entire spectrum as opposed to comparing dots. But much progress is being made in automating such analyses (Clowes, Cooke & Beard 1984, Hewett *et al.* 1985).

Keeping in mind that the main criterion of a successful survey is a high success rate in finding quasars instead of confusion stars, another quasar characteristic has been applied. This is the tendency to variability, which characterizes a minority of quasars (Barbieri & Romano 1981, Pica *et al.* 1980). A survey for blue variable objects should detect only quasars, even though it would not detect all quasars in the fields (Hawkins 1983). Such a survey is especially sensitive to the blazars, or optically violent variables, which are quasar analogs to BL Lac objects. It appears that the relative number of quasars in this category is small (Impey & Brand 1982), even though their detectability is enhanced by beaming effects (Chapter 7). The great promise of surveys for variable quasars is to find enough that it will be possible to learn how quasar properties differ as a function of their variability. This is necessary to study the variability models properly, including possible effects of gravitational lensing (Chapter 4).

2.1.3   Summary of quasar numbers

All optical surveys can be compared and assembled to give summary statistics of just how many quasars there are in the sky. Numbers from survey to survey are adequately consistent that a believable relation can be derived for quasar numbers as a function of magnitude (Braccesi 1983). To almost 20 mag, this is $\log N = 2.2[(B - 18.3)/2.5]$, where N is the number of quasars $\deg^{-2}$ brighter than magnitude B. Within this formula lies the key to why there will always be some disagreements among survey results from different observers. The problem is that quasar numbers increase so quickly with magnitude, at least brighter than 20 mag. Note that the increase is a factor of 7.6 for each magnitude increase in the survey limit. Calibration of photographic survey plates to precisions better than 0.2 mag is not, in my opinion, possible. Variations in sensitivity over the plate, the usual S/N problems on faint objects, and the uncertainties of density to intensity transformations in the photographic process, all combine to produce basic limits on just how good photographic photometry can be. Now notice the consequence. A change of 0.2 mag in the survey limit yields a change by a factor of 1.5 in the number of quasars found! So do not be harsh with observers whose results have large error bars.

Even with a perfect detector, perfect skies, and perfect astronomer, another indeterminate source of uncertainty is also present. This is the obscuration of quasars by dust in our Galaxy, where obscuration varies as a function of position. Canonical formulae exist to take out this 'galactic extinction' as a function of position in the sky (Burstein & Heiles 1978). These results show that extinction can exist for quasars at a level up to a few tenths of a magnitude, introducing an uncertainty in quasar counts comparable to that discussed above. The problem with this absorbing material is that it is distributed inhomogeneously, and increasingly complex relations have been attempted to deal with it. I do not think anyone can honestly state that their extinction corrections are adequate to precision better than 0.1 mag. Enough is just not known about the dust distribution. This is another uncertainty that simply has to be accepted in quasar counts.

The number–magnitude relation above demonstrates why quasars can be found in such large numbers. With wide field Schmidt telescopes, the best equipped being the UK Schmidt in Australia, detections to 20 mag can be achieved in a single exposure, even with spectroscopic survey techniques. If 30 quasars deg^{-2} show up, and the field covers more than 30 deg^2, there is a total quasar presence close to 1000! While these quasars are there, some-where, on the plate, locating them and informing the world is a very tedious task. To have adequate motivation for such surveys, it is necessary to have something in mind other than simply counting quasars.

One category that is still needed, for example, is samples within well defined redshift ranges, especially at the very highest redshifts where knowledge from color based surveys is so incomplete. Also, samples defined for some peculiar but interesting property, such as broad absorption lines, are useful to increase the pool of candidates (Hazard *et al.* 1984). Probably the most pressing need is for sufficient detections of quasars in the range 2.5 $< z < 4$ that the seeming 'turn-on' of the quasar phenomenon can be traced (Chapter 6). Several efforts have to be continued here (Osmer 1982). The wide field telescopes must rely on photographic detection. The necessity is to follow Ly α to redshift 4, or to wavelength 6000 Å. Unfortunately, the existing Kodak IIIaF emulsion – the one needed for sensitivity much beyond

5000 Å – is very non-uniform in spectral response, so that spectral features due to emission lines are hard to distinguish from those due to Kodak. Smoothing out the detector spectral response is the great benefit of CCD imaging chips; but these chips are small, so only a few square arcminutes can be imaged in a single exposure, while piles of magnetic tape are generated whose subsequent analysis is far more time consuming than the actual observing. These two approaches can supplement one another. As it is not obvious which is better, both should continue in the hope of eventually accumulating satisfactory statistics on the quasar redshift distribution above $z = 2.5$.

Note from Table 2.4 (see later) that 41% of quasars to 22 mag are expected to be in the interval $2.5 < z < 5$, unless their evolutionary characteristics for $z > 2.5$ are different than for $z < 2.5$. Learning whether they are is the point of the exercise, of course. It may be expected, therefore, that the most efficient test is simply to accumulate samples of faint quasars by finding objects with blue continua and patiently observing them afterwards one by one to obtain redshifts. It is known that most quasars show substantial absorption blueward of Ly α. This, together with the confusing effects of strong Ly α emission, mean that high redshift quasars may not show distinctive colors unless observed in many bands. Yet, spectroscopic surveys cannot probe as faint as color surveys. As a result, there is motivation for utilizing techniques other than optical, especially if searching for quasars at the highest redshifts. It is noteworthy that, despite the prolific numbers of optically discovered quasars, the quasar of highest known redshift has always been one initially identified as a radio source (Peterson *et al.* 1982).

2.2   Surveys at other wavebands
2.2.1   *Radio surveys*
Even though the optical techniques described are the easiest way to find quasars, it has been extremely important to have the results of other kinds of surveys. All statements that optically discovered quasars are like those discovered any other way can be made only after the fact. It was necessary to have those other surveys before this conclusion could be reached. Initially, of course, quasars as a class were found with radio telescopes, although quasars are not invariably the dominant sources in radio surveys. If by quasar one means the quasar core, or nucleus, detected in an optical observation, these are very rarely strong radio sources. Many radio sources arise from extended lobes at great distance from the nucleus, but the nucleus in such cases may or may not show indications of activity. Such extended sources are often associated with visible elliptical galaxies,

the 'radio galaxies', even though the galaxies do not necessarily contain visible non-stellar activity. Many unidentified extended radio sources exist and are attributed to radio galaxies at sufficient distance that the galaxies themselves are not seen. This is especially characteristic of radio surveys at longer wavelengths because the steep spectra of the optically thin, extended sources are brighter at longer wavelengths (Chapter 7). As radio surveys were done at progressively shorter wavelengths, more compact, unresolved sources were seen which could be identified with quasar cores (Wall 1983). These are the optically thick, 'flat spectrum' sources. But the dominant sources in the faintest surveys now being done are galaxies of one kind or another. Even spiral galaxies become important radio sources at very faint flux limits (Condon 1984).

By observing optically discovered quasars at radio wavelengths, some limited statistical data become available as to the fraction of quasars that are radio sources at a particular ratio of radio to optical flux. This is discussed in terms of the parameter R (Chapter 7), and the flux comparisons automatically refer to the same quasar core because the radio and optical sources are not considered related unless they have the same position. It can be seen from Figure 7.1 that the distribution of fractional detections $G(R)$ is still uncertain. We have no idea whether it is a function of either luminosity or redshift, both of which are important questions to answer. Improving $G(R)$ is a major target of continuing quasar surveys. But with what is available, the $G(R)$ could be folded into the optical quasar distribution in Figure 2.1 to predict the number of quasars detectable to a given radio flux limit that should also be brighter than an optical magnitude limit. For example, an 18 mag quasar would, with the definitions in Chapter 7, have $\log R \sim 0.6$ if the 6 cm radio flux were 1 mJy. From Figure 7.1, only 15% of optical quasars of this mag are radio sources this bright. Such a prediction gives an indication of what we might look for. The faint limit of the existing optical counts in Figure 2.1 is 22 mag; the limiting radio survey flux achievable is ~ 0.1 mJy, which would yield $\log R = 1.2$. From this estimation, we would not expect to identify more than $\sim 10\%$ of the faintest optical quasars as detectable radio sources. Surveys of faint radio sources confirm this.

Radio source survey techniques with the VLA have reached impressive limits, and probably will not be pushed much fainter in the near future. It has been possible to detect sources over wide fields to 0.03 mJy (Condon & Mitchell 1984), but the problem that then arises is in determining what those sources are. Careful comparisons with optical imagery of the same fields are necessary to find quasar candidates; there is no specific indicator of a quasar that can be used with confidence in the radio. From knowledge of

$G(R)$, it is not expected that the majority of faint, optically discoverable quasars in any survey would be found in the radio. Surveys identifying faint radio and faint optical sources in the same fields confirm that the fraction of coincidental detections is small. One survey, for example, pushed radio identifications at 20 cm to 0.6 mJy, and compared these with optical identifications to 23 mag (Windhorst, Kron & Koo 1984). Out of the radio source density of 55 sources deg^{-2} only 10% were identifiable with optical quasars to 23 mag, which is only 3.5% of the total number of optical quasars visible to that magnitude limit (Figure 2.1). Most of the radio sources are faint radio galaxies, distant but not at the high redshifts where the bulk of quasars at this optical magnitude limit are seen. The deepest radio surveys have not been thoroughly compared with optical identifications yet, but that is a challenging area for future work. It took many years just to complete identifications of 3C and 4C radio sources (Wills & Lynds 1978), which are orders of magnitude more radio bright than the faintest sources now detectable.

Figure 2.1. Number of quasars deg^{-2}, n, brighter than observed blue magnitude B. Crosses: counts of quasars discovered by color summarized by Braccesi (1983), with faintest point at 22.5 from survey in selected area 57 by Koo & Kron (1982). Open circle: results from objective prism survey by Savage *et al.* (1984). Solid curve: predicted counts from luminosity function and evolution described in this book, showing only counts for quasars with redshifts ≤ 2.25 so that Ly α emission does not confuse color criterion. This curve was derived independently from the data points shown. Dashed curve: counts of galaxies in same units from Kirshner, Oemler & Schechter (1979) and Tyson & Jarvis (1979).

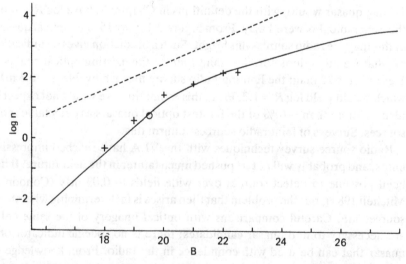

It is not likely that an increased fraction of quasar identifications will be found by pushing radio surveys even deeper, because existing deep surveys (Condon 1984, Windhorst *et al.* 1985) reveal a population of faint, relatively nearby, star-forming galaxies which dominate the counts of faint radio sources. This indicates a basic source of confusion in using radio source counts to study categories of objects (Wall 1983). Much early work was done, leading to conclusions about the evolution of bright radio sources based upon counting numbers of objects as a function of the survey flux limit ($\log N - \log S$ results). To interpret the results, it is necessary to distinguish among the different populations of sources being considered. If radio galaxies, quasars and blue galaxies are all mixed in the counts, little can be learned about the behavior of only one of these components. There is no up-and-coming new generation of radio telescopes to help with this problem. It can only be disentangled by using multi-frequency observations to distinguish populations with some kind of spectral indicators, and calibrating such indicators will require a lot more work in comparing optical and radio sources. The only way to proceed is to make comparisons of optically unresolved objects for positional coincidence with the radio sources and then consider these as radio discovered quasars. One is still left with the same difficulty as in BSO surveys that nothing more can be done until individual spectra of the candidates are obtained. It is the laborious requirements for the latter that have prevented an adequate understanding of the true content of faint radio surveys. Many such surveys exist; hopefully the tools will someday be available to obtain optical spectra of enough faint candidates to understand what the radio surveys are really showing.

2.2.2   X-ray surveys

At the other extreme of the spectrum, X-ray astronomers have been contributing their candidates to the pile. These source lists are not as intimidating to the spectroscopist as the radio and optical samples, because there are not nearly so many of them. The X-ray surveys are dominated by quasars, but only at a surface density of almost 1 deg^{-2}, and only about 100 deg^2 were surveyed during the lifetime of the Einstein Telescope (Gioia *et al.* 1984, Margon, Downes & Chanan 1985). We are statistically in an acceptable situation regarding the optically determined properties of X-ray discovered samples. As a rule, the X-ray surveys revealed quasars similar to those in optical BSO samples. A minority of the X-ray quasars are redder color active galactic nuclei (Stocke *et al.* 1983), but these are not significantly different from some emission line candidates found in optical spectroscopic surveys.

Nevertheless, there were some surprises. Most noteworthy was the tendency for X-ray discovered samples to show systematically lower redshifts than optically discovered samples to the same optical identification limit (Reichert et al. 1982). This has been explained, probably correctly and at least consistently, by allowing quasars to have an optical to X-ray flux ratio that increases with quasar luminosity (Avni & Tananbaum 1982, Kriss & Canizares 1985). That is, more luminous quasars are relatively weaker in the X-ray. Because higher redshift quasars must be more luminous than lower redshift ones at the same magnitude limit, the consequence is that higher redshift examples are harder to find in an X-ray compared with an optical search. The relation required to explain the survey results is that $L_x \propto L_{opt}^{0.7}$ (parameters defined in Chapter 7). If this luminosity scaling is applied to all quasars, optical and X-ray surveys are in agreement both for quasar counts and evolution. Also, this scaling can be explained with at least one physical quasar model which describes the intrinsic energy production in these bands (Tucker 1983). One alternative explanation is that optical obscuration by dust clouds diminishes as a quasar's luminosity increases, such that relatively more optical continuum can emerge for luminous quasars. Another conceivable explanation is that the correlation is just redshift dependent, rather than really luminosity dependent, and arises because the X-ray and/or optical continua have different shapes than those that have been assumed.

In any case, pondering such correlations illustrates what one hopes to find by comparing quasars discovered with widely different techniques. By searching for correlations in the properties, something can be learned about the intrinsic nature of quasars. Equally important in the early stages of understanding is the demonstration that large categories of quasars are not being overlooked because of some severe selection effect in one wavelength regime. To me, this was the most comforting conclusion of the X-ray surveys. The great majority of quasars found have similar emission line and continuum properties as the optically discovered samples, although X-ray samples have led to the discovery of important subsets of active galactic nuclei (Chapter 9). Care must be exercised when comparing survey results in different spectral regions to determine if the same objects are equally represented. It remains the case that quasars seen with one technique are often not seen with another, so proper statistical comparisons of samples require using 'censored data' (Avni et al. 1980, Feigelson & Nelson 1985).

2.3   Distribution of quasars on the sky
2.3.1   Searches for evidence of anomalous redshifts
All numbers given so far have referred only to the average surface density of quasars on the sky. This first-order result is useful for the many

calculations that assume quasars are homogeneously distributed over the sky. Whether this is really the case is a question of importance for several reasons. The first relevant analyses arose as a consequence of questioning the cosmological nature of quasar redshifts.

When the redshift controversy peaked for quasars, empirical evidence favoring non-cosmological redshifts was sought by looking for quasars that seemed to be associated with galaxies at lower redshifts. It was contended that quasars are found close to bright, lower redshift galaxies more often than expected by chance (Burbidge 1979). These arguments are strictly statistical in nature and easy to check, in the sense that the necessary calculations are straightforward. What has not been easy is that the curious features of quasar configurations are pointed out after the fact, without predictive power. The sequence of events has been something like this: a curious fact is discovered within a given quasar sample; perhaps that quasars in a given magnitude range lie closer than expected to peculiar galaxies. That observation cannot produce a meaningful hypothesis unless it can predict a result that will be found using a new, independent sample of quasars. Producing such a sample is tedious and involves a lot of work at the telescope, but somebody eventually does it. Upon reanalysis, the initial curious fact is not confirmed. 'That's unfortunate', comes the response, 'but notice there is a different curious fact in this new sample . . .'. So a new curiosity has to be tested, requiring yet another new sample. It can be readily appreciated why this becomes a frustrating game. Examples of such sequences of events can be followed through Burbidge *et al.* (1971) to Burbidge, O'Dell & Strittmatter (1972); or Burbidge (1979) to Weedman (1980) to Sulentic (1981); or Arp (1981) to Zuiderwijk & de Ruiter (1983) to Arp (1983).

Evidence of various sorts is described in Chapter 4 concerning the cosmological nature of quasar redshifts. It is useful here, however, to point out a simple comparison to test the extreme hypothesis that all quasars are physically associated with nearby galaxies. This test does not depend on finding or testing for the particular galaxy that 'claims' a given quasar. It simply checks whether quasars are distributed with volume in the same way as galaxies. Then, the number of quasars seen as a function of magnitude should increase in the same way as the number of galaxies seen as a function of magnitude. The quasar counts in Figure 2.1 show that quasar numbers actually increase much more rapidly with magnitude of the count limit than do the galaxies in the same figure. Within the cosmological redshift scenario, the quasar counts are explainable by quasar evolution (Chapter 6), which has a very different form from galaxy evolution. Any explanation in the non-cosmological scenario requires an *ad hoc* modification of the initial hypothesis, such as the suggestion that only a minority of quasars have non-cosmological redshifts or are associated with only a subset of galaxies at lower redshift. The same test, incidentally, also demonstrates that most quasars are not gravitationally lensed by intervening galaxies.

Other than curious, improbable configurations noted *a posteriori*, the strongest (i.e. most publicized) case for a non-cosmological redshift has been the close configuration of NGC 4319 – Markarian 205, having redshifts of 0.006 and 0.07 but apparently connected in very deep photographs. Having been the one to point out the redshift discrepancy, I feel some responsibility for what has been made of this pair, and I think that has been way too much. It has been observed so intently and with such heroic techniques that things have been seen which are not commonly seen in pictures of galaxies. Various filamentary features are associated with NGC 4319, and it has been contended that some of these features must be physical connections with the higher redshift Markarian 205. But there are various adequate and equally reasonable explanations of the image properties that do not require any physical association between these objects of different redshift (Cecil & Stockton 1985). Markarian 205 has been fun to challenge people with, but until other remotely similar examples are found – and their absence is not for want of trying – it is useless as a demonstration of non-cosmological redshifts. Any meaningful evidence that non-cosmological redshifts exist, for any or all quasars, has to be argued from statistical comparisons of positions in the sky for quasars and galaxies.

2.3.2   Tests for spatial correlations

There are several important reasons for being able to test for meaningful correlations between the positions of two categories of objects, or among the positions of objects in a single category. The early claims of association between bright galaxies and quasars, in search of proof that quasar redshifts are non-cosmological, were only one example. Now, with the realization of the existence of gravitational lensing, similar associations are being searched for to determine if any dark objects in galactic haloes are intensifying quasar images (Chapter 4). Finally, it is important to determine if quasars cluster among themselves in ways similar to the clustering of galaxies, especially at high redshifts.

Three techniques will be discussed for making these statistical checks. The first and simplest test starts with one category of objects, the one containing the fewer members, and it is determined whether the larger number of objects in a second category in any way associates with the first. This is the procedure, for example, to test whether faint quasars, found in large numbers, are preferentially near bright galaxies, found in small numbers. If a sample containing G galaxies in a total area A is examined, then the total number of quasars – for a random, unassociated quasar distribution – found in area a surrounding the galaxies should be the fraction a/A of the total number of quasars in area A. That is, the number of quasars $n(r)$, expected by chance to be found within a distance r on the sky of any of the galaxies, is $n(r) = \pi r^2 GQA^{-1}$. Here Q is the total number of quasars expected by chance in the area A, so it depends on the surface density adopted for quasars from Figure 2.1. Once a quasar surface density is adopted, to the relevant quasar magnitude limit searched around the galaxies, it is not necessary to pick galaxies within a given area A; the product QA^{-1} is just the number of quasars per unit area, i.e. the surface density in units of Figure 2.1. So one can pick as many bright galaxies as desired, count quasars as a function of r to any desired r, and evaluate the results for the total sample.

An illustration of this is now given using real data. Galaxies used are the 26 in the complete sample of main and companion galaxies Arp (1981) defines (extending from NGC 2549 through NGC 3184 in his Table 1). He has found, by searching, 36 quasars around these galaxies (counting nine near NGC 2639 in another paper). I have tabulated the number of these quasars as a function of distance from their associated galaxy. A basic uncertainty in such counts is the true limiting magnitude of the survey, so I have separately listed all quasars with $m < 20$ and all with $m < 19$ to accommodate this uncertainty. To magnitude 19, the observed surface density of quasars in the sky, from Figure 2.1, is $QA^{-1} = 4.1 \text{ deg}^{-2}$ and to magnitude 20 is 31 deg^{-2}. (These are actually densities for quasars found as blue stellar objects, but that is what Arp also searched for.) In Table 2.1, the numbers of quasars found are compared with the number expected by chance using these results with the relation above.

Notice the problem is interpreting the results. The objective was to test if quasars are found close to bright galaxies more often than expected by

Table 2.1. *Quasars near bright galaxies*

	2′	4′
Radius of search		
Number of quasars with $m < 20$ expected for random distribution	2.8 ± 1.7	11 ± 3.3
Number of quasars with $m < 20$ found by Arp	5	11
Number of quasars with $m < 19$ expected for random distribution	0.4 ± 0.6	1.5 ± 1.2
Number of quasars with $m < 19$ found by Arp	1	5

chance. The greatest 'discrepancy' occurs for $r < 4'$ and $m < 19$, where five quasars were found versus 1.5 expected. This does not mean that quasars are three times as prevalent as expected within $4'$ of a galaxy! The numbers are small. The significance of the results has to be evaluated in context of the basic statistical uncertainty of any observation. If a number N is the true value that would be found with unlimited measurements, we know only that there is a 68% chance that the observed value of N with a single measurement is within the range $N \pm N^{0.5}$, or within one standard deviation (σ) of the real value. Within two standard deviations, the chance is 95%. Put another way, there is always a 32% chance that the measured value will lie outside the 1σ range about the real value. So if we looked at three bins, we would *expect* to find one exceeding 1σ. In this example, there are four bins examined, which gives four chances to find a result differing by more than 1σ from expectation; it is no surprise that one is found. That particular bin tells us nothing meaningful. Only if a completely separate and independent sample were evaluated which also showed a deviation in that *same* bin would there be evidence for an unusual property of that bin.

Extensions of this 'root N' concept of statistical uncertainty have been comprehensively produced by statisticians to provide much more sophisticated tests of confidence in a result. Such checks as chi-squared tests and Kolmogorov–Smirnov tests are expansions upon this basic concept of statistical uncertainty. All one gets in the end is a probability. It is up to the observer to obtain enough data that these probabilities, for or against a given hypothesis, become believably large.

The method just described requires counting all quasars within a radius r of each galaxy, extending out to the maximum r of interest. When the number of galaxies in the test is large, this can be tedious. Also, problems may arise with overlapping fields around different galaxies, so those would have to be corrected. A much more efficient way to obtain the data would be simply to measure the distance r to the nearest quasar from any galaxy. Obviously, such measurements could easily be made when intercomparing any two samples of objects. This is the technique of the 'nearest-neighbor' test, which can provide a test of equal utility and easier implementation (Zuiderwijk & de Ruiter 1983). If an object, quasar or galaxy, is chosen as a test point, any class of objects randomly distributed about it should have a poissonian distribution. Then, the probability of finding one member of the class at a distance r to $r + dr$ from the test point is $P(r) = 2\pi r N \exp(-\pi N r^2)$ dr, where N is the surface density of the presumed random class of objects. Equivalently, the actual number of nearest-neighbors that should be found within distance $r + dr$ is $P(r)T$, where T is the total number of nearest-neighbors measured. In practice, one wishes only to find the nearest-neighbor of a given test point, then go to another test point and find its nearest-neighbor, etc. The results can be expressed in terms of the fraction of nearest-neighbor distances smaller than x. For the random distribution, this fraction should be $F(x) = 1 - \exp(-\pi Nx^2)$. The expectations of $F(x)$ can be compared with what is actually found, keeping in mind the same cautions about root n statistics discussed above when applied to observed numbers.

This nearest-neighbor test is now illustrated using an all-sky sample of quasars – the Palomar Bright Quasar Survey (PBQS) (Schmidt & Green 1983). We wish to test if these quasars are preferentially associated with galaxies, as would be the case if they are brightened by lensing from galactic objects. Implicit in this test is the requirement that the quasars be far beyond the galaxies, rather than at the same distance, so we should restrict the comparison to high redshift quasars and low redshift galaxies. To do this, quasars in this sample with $z < 0.1$ are not included, nor are the few quasars at slightly higher redshifts that show a galactic envelope in photographs. These criteria omit 35 of the 114 quasars in the PBQS. The remainder are compared in position with the bright galaxies cataloged by Zwicky (necessitating the further omission of five quasars below the celestial equator, in an area of the sky not covered by Zwicky's catalogs). Table 2.2 contains the quasars and the distance to the nearest Zwicky galaxy, as measured on the Palomar Observatory–National Geographic Society Sky Survey (these unpublished measurements were made by K. C. Phillips). Using these results, a nearest-neighbor distribution is shown in Table 2.3. Bin size in any test like this is arbitrary; one wishes small bins to show fine-scale effects, but the bins need to contain several objects to reduce the root n uncertainties. The 1σ uncertainties are also shown for each bin in Table 2.3, so the observed distribution is what is compared with the expected distribution, with uncertainties, from the relation above. Using a galaxy surface density of 2 deg^{-2} (18435 Zwicky galaxies in the 9000 deg^2 of this survey), the expectations become those compared in Table 2.3, for $T = 74$, and $N = 5.6 \times 10^{-4}$ arcmin^{-2}. (Because dr is comparable to r for many of the bins, the product $2\pi r$dr is equated to the actual area of the bin, and the r which enters $\exp(-\pi Nr^2)$ is taken as the midpoint of the bin.)

The smallest separation analyzed is the 3' to 10' bin, just because there is no quasar closer than 3'. The vagaries of statistical samples are evident in the results. None of the closest three bins to quasars show an excess of galaxies compared with the number expected for a random distribution, with third and fourth bins showing a deficiency and then an excess. The conclusion from this analysis has to be that there is no evidence in this sample for the

association of bright quasars and galaxies having different redshifts, which
was the hypothesis under test. (Because the low redshift quasars or those
with envelopes were excluded, the result does not address the question of
quasars being found in clusters of faint galaxies at the same redshift as the
quasar.) To go any further and search for any associations that might be
present in relatively few cases, we must overcome the uncertainties from
statistics of small numbers. The sample of quasars to be tested would have to
be increased. Because the number of detectable quasars increases by more
than a factor of seven per one magnitude increase in the survey limit
(Figure 2.1), there are plenty of quasars to be found. Wide field surveys with
automated detection techniques can easily push to 19 mag, so the numbers

Table 2.2. *Quasars from PBQS with nearest galaxy*

Quasar[a]	Closest galaxy (arcmin)	Quasar	Closest galaxy (arcmin)	Quasar	Closest galaxy (arcmin)
0003 + 158	32.5	1206 + 459	35.8	1543 + 489	31.4
0026 + 129	19.0	1216 + 009	11.2	1545 + 210	12.3
0043 + 039	38.1	1222 + 228	21.3	1552 + 085	25.8
0044 + 030	49.3	1226 + 023	51.5	1612 + 261	16.8
0052 + 251	23.5	1241 + 176	71.7	1613 + 658	17.9
0117 + 213	4.5	1247 + 267	43.7	1617 + 175	50.4
0157 + 001	15.7	1248 + 401	56.0	1630 + 377	19.0
0804 + 761	24.6	1254 + 047	10.1	1634 + 706	43.7
0923 + 201	9.0	1259 + 593	6.7	1700 + 518	33.6
0946 + 301	37.0	1307 + 085	12.3	1704 + 608	11.2
0947 + 396	34.7	1309 + 355	23.5	1715 + 535	37.0
0953 + 414	12.3	1322 + 659	68.3	1718 + 481	22.4
1001 + 054	11.2	1329 + 412	39.2	1114 + 445	67.2
1004 + 130	10.1	1333 + 176	11.2	1115 + 080	7.8
1008 + 133	33.6	1338 + 418	33.6	1116 + 215	15.7
1012 + 008	15.7	1402 + 261	40.3	1352 + 011	35.8
1048 + 342	16.8	1407 + 265	6.7	1352 + 183	12.3
1049 − 005	13.4	1415 + 451	37.0	1354 + 213	31.4
1100 + 772	14.6	1425 + 267	9.0	2112 + 059	40.3
1103 − 006	20.2	1427 + 480	16.8	2233 + 134	3.4
1121 + 422	54.9	1444 + 401	53.8	2251 + 113	9.0
1138 + 040	31.4	1512 + 370	33.6	2302 + 029	6.7
1148 + 549	33.6	1519 + 226	41.4	2308 + 098	35.8
1151 + 117	31.4	1522 + 101	20.2	2344 + 092	16.8
1202 + 281	16.8	1538 + 477	35.8		

[a] Quasar names follow the standard convention based on their position; that is, 0003
+ 158 is located at right ascension 00 hours, 03 minutes and declination 15.8 degrees.

within bins like those in Table 2.3 could be increased by factors of several hundred using quite feasible observations. This kind of statistical association test is a good use of such surveys, because all that is needed are positions for large numbers of quasars known to have redshifts above 0.1. Neither their precise magnitudes nor their precise redshifts are required.

The attraction of the nearest-neighbor test is that it only requires the comparison of the position of one object with that of one other object. It might be suspected that more information would be contained by comparing the position of a given test point with that of every other point in the sample of interest. This is the basis of the angular covariance function, or two-point correlation function (Peebles 1973). In effect, all objects in a sample are correlated in position with all other objects, or two different samples are similarly compared, to search for any scale size on which the distribution is non-random. This is the basic test for clustering of galaxies (Seldner & Peebles 1979). It can be applied as well to quasars. Of course, a true test for clustering requires that objects be associated in three dimensions, not just the two dimensions seen on the sky. Such a three-point correlation function can be produced only for objects with known distances, and even then the function becomes quite complex when cosmological relations for distance have to be used. Only the two-point function is discussed below, but it would be quite useful for quasars within a narrow shell of distance, i.e. within a small redshift range. Samples of such quasars are being accumulated with existing survey techniques even if their precise redshifts are not known.

Table 2.3. *Nearest-neighbor quasar – galaxy test*

Separation	Galaxies expected[a]	Galaxies found
3' − 10'	11 ± 3.3	9
10' − 20'	26 ± 5	23
20' − 30'	22 ± 5	8
30' − 40'	10.6 ± 3.2	20
40' − 50'	3.3 ± 1.8	6
50' − 60'	<1	5
60' − 70'	<1	2
70' − 80'	<1	1

[a] The 1σ uncertainties are derived as (number expected)$^{0.5}$, so there is a 68% chance that a measured number should fall in this range. When the number expected falls below unity, so that fewer than one event is involved, this measure of statistical uncertainty has no meaning.

Application of the two-point correlation test to such samples (Savage *et al.* 1984) can give valuable information on whether quasars show similar clustering properties to galaxies, but results to date find no evidence of such clustering.

The two-point correlation function for objects within a sample is $w(\theta)$, which gives a measure of the fractional enhancement or deficiency for the density of test points at any scale θ compared with their overall mean density. This is defined by $N(\theta) = N_m[1 + w(\theta)]A$. Here, N_m is the mean surface density (number per unit area) of objects in the sample. $N(\theta)$ is the number of objects found in total area A located within distances θ to $\theta + d\theta$ from single test objects. This total area is the product of the area examined per test object multiplied by the number of test objects used. For a random (poissonian) distribution, the $w(\theta)$ should be zero at all scales θ. That is, the number of objects in any area is just $N_m A$, regardless of the position of A. But if there is any angular scale over which objects tend to cluster, w is larger than zero at scales inside this clustering length scale. If there is a scale at which the objects avoid each other (anti-clustering), w is negative within that scale. To calculate $w(\theta)$, all points in the sample have to be considered as test points, and $N(\theta)$ calculated for all other points about each test point. This is equivalent to measuring $N(\theta)$ as the total number of pairs within the entire sample having separations between θ and $\theta + d\theta$. Each separation is counted twice so each object pair counts twice in $N(\theta)$. The total area A examined to find this $N(\theta)$ would be $2\pi T\theta d\theta$, with T the total number of test points in the sample and $2\pi\theta d\theta$ the area examined for each point.

Similarly, two separate populations of objects can be tested, such as comparing quasars with galaxies. The two-point correlation function $w_{ab}(\theta)$ between the two populations a and b is defined by $N_b(\theta) = N_{mb}[1 + w_{ab}(\theta)]A$. Now, $N_b(\theta)$ is the number of objects from population b located in area A at distances θ to $\theta + d\theta$ from members of population a. N_{mb} is the mean surface density of population b. That is, $N_b(\theta)$ is the total number of objects in population b located at distances θ to $\theta + d\theta$ from *any* member of population a.

It can be seen that use of these two-point correlation functions requires major number crunching, because the position of every object has to be compared with the positions of all other objects. Even when the crunching is finished, the results are not interpretable unless the statistical uncertainties involved are considered. These are not as straightforward as in the earlier tests discussed. It is convenient to rewrite the above relations as $w(\theta) = [N(\theta)/N_m T 2\pi\theta d\theta] - 1$, with $N(\theta)$, T and N_m all measured as described above. The objective is to determine $w(\theta)$ for each scale length θ, and then note whether $w(\theta)$ deviates from zero at any scale. Error bars have to be applied to determine if any deviations seen are statistically meaningful. (This is a separate check from whether observational selection effects, such as choice of observing area, may have influenced $w(\theta)$.) Statistical uncertainties $\Delta w(\theta)$ are approximately $\Delta w(\theta) = [1 + w(\theta)]^{0.5}[N(\theta)/2]^{0.5}$ if the points are not strongly clustered. (If they are strongly clustered, there are fewer 'independent' pairs, so the $\Delta w(\theta)$ increases for the same number of points.) Reducing $N(\theta)$ by the factor 2 yields the number of true pairs, each pair having been counted twice to get $N(\theta)$. In comparing two populations to determine $w_{ab}(\theta)$, each member of b is counted only once per pair, so $\Delta w_{ab}(\theta) = [1 + w_{ab}(\theta)]^{0.5}[N_b(\theta)]^{0.5}$. As is the case in many astronomical discoveries, the most believable results come when a correlation scale indicated by one sample is confirmed with a completely independent sample.

The statistical tests described are unambiguous. Anyone can use them for any sample. The only requirement for obtaining meaningful scientific results is to define *before testing* what correlations are to be tested with the sample available. Quasar surveys are continuing and there is really no limit to the number that can be found. Applying the tests described without advanced knowledge of the location of the quasars has been done for several samples, none of which show evidence for association of high redshift quasars with low redshift galaxies (He *et al.* 1984, Shanks *et al.* 1983). This impacts on hypotheses of both non-cosmological redshifts and gravitational lensing. If the galaxy samples tested become so faint that the galaxies are comparable in redshift to the quasars, the cosmological redshift hypothesis would predict associations as quasars appear within clusters of faint galaxies. Such associations of quasars with galaxies at the same redshift are observed (Yee & Green 1984, Heckman *et al.* 1984). There is much utility for statistical tests in astronomy. Confusion arises only when all users do not play with the same rules.

My conclusion from the analyses summarized is that the 'evidence' for non-cosmological redshifts is a collection of unrelated curiosities having no predictive power. Because of the reasons summarized in this chapter and Chapter 4, I am convinced that quasars have cosmological redshifts, i.e. redshifts that relate to distance the same as for other mass in the Universe. (Knowing the form of this relation still requires determining the correct cosmology, as cautioned in Chapter 3.) Furthermore, I do not believe that any of the curiosities found to date provide acceptable evidence for non-cosmological redshifts even for a small minority of quasars. The problem, of course, is that if there were indeed a few quasars manifesting non-cosmological redshifts in a variety of peculiar and non-reproducible ways, no unique curiosity will ever be sufficient to prove this. In summary, I do not believe the prospect of finding non-cosmological redshifts to be sufficiently promising that more effort should be invested in the task. Such an approach is not self-blinding, because the opportunity for serendipitous discovery of curiosities is always available in quasar surveys undertaken for any other purpose.

2.4   Promise of future surveys

We are on the threshold of being able to increase enormously the survey capabilities in optical, infrared and X-ray astronomy. These great increases will come about with a new generation of telescopes in space, and the ability to handle quantitatively the immense production of ground based optical survey telescopes. With the potential of these instruments, we need to know what is expected and what would be unexpected. We need the signals to tell about a new breakthrough in awareness of quasars. So I have taken the synthesis of what we know now, as expressed in the luminosity function in Chapter 5, the evolution in Chapter 6, and the continuum properties in Chapter 7, to predict what would be expected with the new instruments. Predicting the results of surveys going deeper than those done now requires many assumptions about the flux distributions of quasars in the sky. Such assumptions depend critically on the form of evolution assumed, so the predictions provide nothing more than a benchmark for testing if the evolution parameters are reasonable. Nevertheless, it is useful to have some guide as to what to anticipate when probing to unexplored flux levels. Existing data yield a consistent picture, based on the assumption of pure luminosity evolution extending from the local quasar luminosity function back to a redshift of about 2.5. Beyond that redshift is the most exciting regime to probe with new observations, because at some epoch the quasars must have formed. There are hints from optical data that the epoch at which the character of evolution changes may be observable in the vicinity of $z \sim$ 2.5. For now, I provide a simple scaling incorporating no quasar 'turn-on'; real observations can be compared with this scaling to search for the quasar formation epoch. All calculations adopt the cosmological relations for $q_0 =$ 0.1.

This scaling begins by taking existing surveys, describing their limits, and summarizing the characteristics of the quasar populations found. These characteristics have to agree in the main with the expectations of the evolution parameterization, of course, because the latter was derived from the real data. A lot of smoothing has been necessary, in terms of spectral shapes and magnitude limits, so the reader should not look for perfect agreement between my summary and any individual survey in the literature. The next step is to imagine the extension of the survey capabilities, extrapolating to new instruments and technologies that should be available within a decade.

Results are incorporated in Tables 2.4, 2.5 and 2.6 giving, for a particular survey type and limit: (a) the total number of quasars to $z = 2$; (b) the number with $2 < z < 2.5$; (c) the number with $2.5 \leqslant z \leqslant 5$; ($d$) the redshift at which the number peaks. The usefulness of these particular characteristics of a survey is: (a) Quasars with redshifts below 2 are recognizable by their optical colors alone as being blue stellar objects. For redshifts much above 2, a sufficiently strong Ly α line can distort this blue color. So color based surveys are confident only at such redshifts. Optical spectroscopic surveys will have high proportions of candidates with only continuous spectra for these redshifts, the strong Ly α line being the one which is easiest to recognize on such spectroscopic surveys. (b) It is in the redshift interval 2–2.5 that wide field spectroscopic surveys are at their best, because both the Ly α and C IV 1550 lines are in the observable spectrum. Even if both lines are not seen on the discovery spectra, subsequent follow ups will be easiest for quasar redshift determinations from this, the strongest line pair in quasar spectra. Similar confirmation bias applies to follow up of X-ray or radio surveys, but as such follow ups are now being done with CCD detectors, having a smooth and efficient spectral response extending to 7000 Å, the feasible redshift interval extends to $z \approx 5$, at which even Ly α would no longer be detectable. This accounts for the listings under (c). (d) To give an approximate idea of what epoch is being probed most efficiently by a given technique, the most probable redshift of any quasar found is given. Comparing this number with any real data will give a first-order indication of the direction in which the evolutionary parameterization errs.

2.4.1 *Faint X-ray, optical and infrared surveys*

Existing X-ray surveys have never been able to determine the spectral properties of any quasar found, nor has it even been possible to be confident that any source found is a quasar. Subsequent optical spectroscopy of identified X-ray sources showed that more than half are quasars, with this fraction increasing as the surveys reach fainter limits. Expectations in Table 2.5 are derived using the representative X-ray parameters now known. The quasar continuous spectrum is assumed to relate between X-ray and optical by the slope $\alpha_{ox} = 1.3$ and the slope of the X-ray continuum itself is taken to be $\alpha = -0.7$. More is said about these parameters, their scatter, the X-ray continuum, and its importance in Chapter 7. Even the survey

Table 2.4. *Quasar redshift distributions for optical surveys*[a]

Optical limit (B mag)	Number deg^{-2} $z < 2.0$	Number deg^{-2} $2.0 \leqslant z < 2.5$	Number deg^{-2} $2.5 \leqslant z \leqslant 5.0$	Most probable z
16	6.2×10^{-3}	0.3×10^{-3}	0.3×10^{-3}	<0.1
19	2.5	0.24	0.37	0.3
22	140	41	124	1.5
25	950	340	1340	1.9
28	3550	1440	6200	2.2

[a] Results predicted using continuum shape, luminosity function and pure luminosity evolution described in this book, *without* adopting any redshift limit for modifying the form of quasar evolution (Chapter 6). This last assumption is unrealistic and probably makes the numbers in the fourth column much too large. This applies also to Tables 2.5 and 2.6.

Table 2.5. *Quasar redshift distributions for X-ray surveys*[a]

Instrument	Number deg^{-2} $z < 2.0$	Number deg^{-2} $2.0 \leqslant z < 2.5$	Number deg^{-2} $2.5 \leqslant z \leqslant 5.0$	Most probable z
Einstein 'deep' survey	190	45	104	1.3
AXAF	1150	360	1200	1.7

[a] See footnote to Table 2.4.

Table 2.6. *Quasar redshift distribution for SIRTF spectrometer*[a]

Infrared flux limit	Number deg^{-2} $z < 2.0$	Number deg^{-2} $2.0 \leqslant z < 2.5$	Number deg^{-2} $2.5 \leqslant z \leqslant 5.0$	Most probable z
0.25 mJy at 10 μm	110	26	50	1.1

[a] See footnote to Table 2.4.

limits are spectral dependent and difficult to define for X-ray results. To interpret results at different redshifts, the X-ray spectrum must be known. Because the detectors that have been used are broad band, a spectrum also has to be assumed to extract a quasar flux. Two levels of surveys for quasars have been published: the 'medium sensitivity survey', covering many fields observed with the imaging proportional counter on the Einstein Observatory (Gioia *et al.* 1984) and a few 'deep surveys', using primarily the high resolution imager on the same satellite (Griffiths *et al.* 1983). In both cases, limiting fluxes are quoted for the entire bandwidth of the detectors, which is approximately between 0.3 and 3.5 keV, the 'soft' X-ray band. This needs to be changed to a monochromatic X-ray flux for consistency with calculations at other spectral regions. The effective energy is approximately 2 keV; to list a monochromatic X-ray flux limit at 2 keV, the limiting fluxes quoted over the entire bandpass are changed to f_ν by assuming the bandpass width to be 3 keV. With this approximation, the existing limit of the medium survey is about 1.3×10^{-31} erg cm^{-2} s^{-1} Hz^{-1} and the deep limit about 1.3×10^{-32}; the latter is in Table 2.5.

Performance parameters for AXAF are not well known at this time. Detectors will be different from those on the Einstein Observatory, and may be most efficient at slightly higher energies. Also, these detectors will obtain much improved spectral information. To normalize calculations, it is assumed that the deepest feasible AXAF exposure which can obtain a continuous spectrum of a quasar would reach a 2 keV limiting flux of 10^{-33} erg cm^{-2} s^{-1} Hz^{-1}.

Folding these parameters into the best estimate of quasar evolution – the pure luminosity evolution model in Chapter 6 for evolution as seen with X-ray samples – yields the results in Table 2.5.

An important constraint on the expectations for faint X-ray surveys is that the total flux from all quasars must not exceed the limits set by the X-ray background. Whether that happens or not with calculations such as those given depends on slight adjustments of the evolutionary and spectral parameters (Piccinotti *et al.* 1982, Maccacaro, Gioia & Stocke 1984). No accurate determination is possible until the true shapes of quasar X-ray continua are known. Understanding the nature of the X-ray background is the most pressing issue in X-ray astronomy. The background is bright, comparable to the combined flux of all discrete sources. There is the possibility that the background is showing a specific and important epoch in the universe, such as that of galaxy formation. This cannot be known until the flux and spectrum of the background is known independently of the quasar contribution. Being able to make this correction is adequate motivation for learning all we can about faint X-ray quasars.

In the optical, there are many results from existing data, but extrapolations are important in context of HST and improved ground based spectroscopic survey techniques. For the calculations presented, the continuous spectrum of quasars in the optical is assumed to have index -0.5 (Chapter 7). Existing optical counts of number of quasars as a function of magnitude, based on their color properties, go close to 23 mag (Koo & Kron 1982). This kind of survey can be done with fine grain photographic emulsions on 4 m-class telescopes. With CCD detectors, such counts might be pushed a couple of magnitudes fainter, but the real gain will come from HST, which may be able to image unresolved sources to 28 mag. Existing and future counts of quasars are summarized in Table 2.4 and Figure 2.1 on these bases. The problem with counting quasars only on a color selection criterion is the uncertainty by contamination from other objects. Obviously, stars which mimic quasar colors could be a problem. Also, faint and compact galaxies could begin to show in significant numbers when searching for faint blue starlike objects. Discriminating against marginally resolved sources, like faint galaxies, can be done using image processing techniques that fit intensity profiles to sources (Tyson & Jarvis 1979). This technique is very computer intensive, but is an important step in improving the counts of faint galaxies and quasars. Careful selection of the color bands utilized can help in screening out stars. The ability to measure large numbers of quasar candidates with machine based techniques will provide the greatest improvement in ground based quasar counts. An ultimate deficiency will nevertheless remain in that these broad band observations cannot give any information on the redshifts of the quasars found.

Only one survey technique, out of all being discussed here, is at present able to yield any information concerning a quasar redshift from the discovery observation. Within certain redshift ranges, and for objects with sufficiently strong emission lines, the low dispersion spectroscopic surveys can give redshifts. This is a survey technique that is likely to benefit immensely from machine based plate measurements. Faint emission lines can be found with more confidence, and quantitative continuum magnitudes can be determined (Clowes *et al.* 1984, Hewett *et al.* 1985). The only reason that broad band surveys can reach fainter magnitudes is because, by dividing the spectrum into several resolution elements, the spectroscopic observation diminishes the quasar flux per unit area of the photograph compared with the background sky brightness. With the spectroscopic technique, there are tradeoffs between using wide field Schmidt telescopes which cover larger areas, and larger telescopes which go fainter in smaller areas. The limit for existing Schmidt surveys is approximately 19 mag and for large telescopes about 22 mag, each of which may be pushed about a magnitude fainter using machine based techniques, or using CCD detectors on existing telescopes. The big hope is imagery through a transmission grating on HST, which could push low dispersion spectroscopic surveys to 25 mag. These kinds of surveys are particularly well suited to the interval $2.0 < z < 2.5$ (Weedman 1985), which is why there are numbers in Table 2.5 for that interval.

For comparison to the optical quasar counts, a useful number for ground based imagery is the overall background density of unresolved sources (faint stars and compact galaxies) on the deepest available photographs. This is about $1.5 \times 10^4 \, \text{deg}^{-2}$ (Windhorst *et al*. 1984), or about ten times the number of quasars expected to 24 mag.

The remaining spectral regimes, infrared and radio, may not prove too helpful as quasar survey techniques because of the difficulty of distinguishing quasars from other faint sources. This problem has already been discussed regarding radio observations. Little survey work for faint extragalactic sources had been done in the infrared until the IRAS survey, and it seems from this result that the majority of extragalactic infrared sources are very reddened galaxies (Soifer *et al*. 1984). The infrared should not be expected to prove a spectral region for prolific quasar counting (Neugebauer *et al*. 1984). On the other hand, obtaining the spectra of quasars in the infrared remains an important goal, and will prove critical if there exist many quasars which are heavily dust obscured. If, for example, quasars at early epochs in the universe were surrounded by dust, they might be invisible as optical sources so that optical counts would give erroneous results regarding the epoch of their development. As a guide, therefore, as to what should be there, quasar count estimates are also given for the infrared. Spectral parameters are known reasonably well for quasars discovered with other techniques, but the population of infrared-dominant quasars is underestimated in such a calculation. Observed quasars are known to have continuous spectra from the optical through the infrared of index approximately -1. If this spectrum is used to make predictions, the infrared does not yield any gains in flux limits over available optical surveys. Only quasars with indices steeper than this will be easier to see in the infrared, and these would characterize any heavily obscured sample. What we really wish in contemplating infrared results is an idea of what constraints are provided on quasars at high redshift that might be hidden optically but observable in the infrared; this is a possible explanation for past difficulties in finding very high redshift quasars optically. Consequently, only the numbers for $2 < z < 5$ are really of interest for Table 2.6. While these are obtained by maintaining a spectral slope of ν^{-1} from the optical to infrared, the results are lower limits for any quasar population for which blue light is absorbed and re-emitted in the infrared.

Existing benchmarks are taken from the capabilities of the IRAS deep limit at 12 μm, adopted as 250 mJy; virtually no high redshift quasars should be seen with this instrument. A major improvement will come with SIRTF, but it is not particularly clear how quasars will be sorted out from other sources even with that instrument. The spectroscopic capabilities are important to estimate in this context, however, in case that candidates for shrouded quasars at high redshift need to be examined. It seems that low resolution spectra should be obtainable to 0.25 mJy with this instrument, to which flux limit the quasar population lower limits should be as in Table 2.6.

As a final evaluation of survey potential, we note from the results in the tables that optical surveys will give the maximum feasible quasar return. At the limit of 26 mag, there should be about 2000 deg^{-2}. Multiplying this by the area of the sky tells us that there are 100 million quasars accessible to human ken. How many of these must be found before we are satisfied with our knowledge of quasars?

3

COSMOLOGY

3.1 Introduction

Decisions have to be made regarding the nature of the universe before observed properties of quasars can be transformed into intrinsic properties. We must adopt a cosmology. My objective is not to derive formally the equations of cosmology, as that has been done thoroughly many times. The ultimate purpose of this chapter is to justify these equations qualitatively while putting most of the quantitative effort into describing how to use them to study quasars. The cosmologies to be adopted were derived long before quasars were discovered, but their use became a much more serious affair once quasars had to be considered. This is because the redshifts of quasars are often sufficiently high that differences among different cosmologies become quite large. For most galaxies, by contrast, even the difference between newtonian and relativistic cosmologies can be ignored. We can certainly not yet guarantee the equations of favored cosmologies as applying to the real universe. Once a single cosmology is adopted for it, the universe is forced to become a simple place as regards the structure of spacetime. As long as such a simple universe fits what we see, it is appropriate to retain it. There is certainly little motivation for arbitrarily postulating increased complications; nevertheless, observers must forever be on the alert for those anomalies which would show the simple models to be valid no longer. There is a vested interest in having a simple universe, as that is the only kind we can understand. One way to keep the universe simple, by definition, is to avoid seeing the anomalies, spending effort only on refining the adopted model.

This danger is avoided as long as we appreciate that no cosmology can be considered correct until it is consistent with all available data. So far, the Friedmann cosmologies have come very close to this requirement. Many sophisticated discussions of such cosmologies are available; some of the basics are reviewed here because of the great importance of the primary inconsistency that now remains. This is an inconsistency between the age of the universe derived cosmologically and that derived from stellar evolution theory. More specifically, it is the debate over the value of the Hubble constant. Any researcher dealing with extragalactic astronomy must have a sufficient understanding of the cosmological fundamentals that the importance of this issue can be appreciated. Primarily for this reason, I review the basic conceptual derivations of the Friedmann cosmologies.

The extent to which some cosmological fundamentals are still fragile can be appreciated by realizing that it has never been *proven* that galactic redshifts are caused by the expansion of the universe as opposed to some process causing photons to lose energy as they traverse space. This 'tired light' hypothesis was seriously considered when the earliest observations of redshift were made, but even at present the dismissal of that hypothesis remains based more on consensus than on experiment. The issue is still viewed much the same as stated by Hubble & Tolman (1935):

> Until further evidence is available, both the present writers wish to express an open mind with respect to the ultimately most satisfactory explanation of the nebular red-shift and, in the presentation of purely observational findings, to use the phrase 'apparent' velocity of recession. They both incline to the opinion, however, that if the red-shift is not due to recessional motion, its explanation will probably involve quite new physical principles.

Despite widespread acceptance of the expansion of the universe as the explanation of all redshifts for galaxies and quasars, there are a few active efforts to argue otherwise. Recent examples of alternative explanations have been the idea of quantized redshifts (Cocke & Tifft 1983) and the chronometric cosmology (Segal 1980). These are not in the same category as efforts to demonstrate 'non-cosmological' redshifts for quasars (Burbidge 1979). The issue in the latter case is whether some quasars have redshifts arising from different mechanisms from the 'cosmological' redshifts accepted for galaxies. Treating these various alternatives, therefore, must be done differently. The issue of non-cosmological quasar redshifts is one that depends strictly on comparisons of data for galaxies and quasars, and is an issue only at the level of statistical arguments. Consequently, it was considered when the distribution of quasars in the sky was discussed in Chapter 2. As emphasized therein, my conclusion, representative of majority opinion, is that the same equations of cosmology apply to quasars as to any other component of the universe. No more will be said about that issue in this chapter's discussion of cosmology. Alternative cosmologies, on the other hand, treat galaxies and quasars the same. Segal, for example, contends that support for the chronometric cosmology is found with the interpretation that this cosmology does not require quasar evolution.

The chronometric cosmology is a modern example of a non-expanding cosmology, one in which redshift arises from something other than a Doppler shift. So it raises again the same fundamental question that worried Hubble. There does now seem to be the chance, with new observational techniques, to test once and for all the presumption that redshifts are caused by the general expansion. This opportunity arises because of the effects of expansion on the observed surface brightnesses of distant galaxies. Surface brightnesses are discussed in Chapter 4, where some differences are mentioned that should be observable if the universe is not expanding. In summary, the result is that the surface brightnesses and the resulting observable diameters would not diminish nearly so quickly with redshift in a non-expanding universe. Appreciating the reasons for this first requires an understanding of the basics of expanding cosmologies.

The issue, therefore, is whether the Friedmann cosmological models built up from observations of galaxies have given a correct picture of the real universe. Two fundamental puzzles remain to be solved before we can be really comfortable with these models. The first, already mentioned, is to account for a seeming inconsistency in the cosmological age of the universe compared with stellar ages derived from theories of stellar evolution. The second is to explain why the actual universe is so close to the boundary between open and closed.

3.2   Fundamentals of a cosmology

A basic simplification is imposed on the universe because of our inability to deal with too complicated a structure. This is that of the 'cosmological principle', which is actually the 'cosmological assumption'. This requires that the universe be homogeneous and isotropic everywhere but not at every time. Any observer anywhere, but not at any time, must see the same thing. Spatially, the universe must be uniform, although it is allowed to change its properties with time. (The steady state cosmology of Hoyle (1949) requires a 'perfect cosmological principle' that does not allow changes even with time; as mentioned in Chapter 6, one of the uses of quasars is to rule out the steady state cosmology, because quasars are seen to evolve in this cosmology.) The great working advantage of this principle is that the behavior of the entire universe mimics the behavior of any one of its bits. That means we can study the local part of the universe, being the only part we can see well, and apply conclusions to the rest of the universe from these local studies. Much of the research that can be called observational cosmology is, in fact, directed toward determining just how large a volume of the universe must be studied to consider it truly representative of the whole.

What we wish to describe with the equations of cosmology is the overall behavior of the substratum within which the particles of the universe are embedded. In a cosmological view, mass particles define locations in space, and the motions of these particles show what is happening to the underlying space itself. If the ensemble of particles expands, that means the space of the universe expands. Because of the cosmological principle, it makes no difference what bit of space is examined, as it behaves like all the other bits, the sum of which describes the universe as a whole. Within a sufficiently small portion of the universe, newtonian physics is adequate to determine the properties of that portion. Then, if that portion were large enough to be a representative sample, we could apply our conclusions to the rest of the universe via the cosmological principle. Unfortunately, no checks on the cosmological principle can be made for a newtonian universe, nor can we inquire of the spatial size of the universe. To do either requires interpreting information carried through large distances by light, and there are no mechanisms within newtonian cosmologies to account for effects which the structure of the universe might have on the propagation of light. So general relativity is needed, and the differences between relativistic cosmologies and newtonian cosmologies become very large for quasars. Consequently, the procedure that follows will be to illustrate the basic concepts of cosmological derivations by working through a newtonian cosmology; then, the correct equations of general relativistic cosmologies will be introduced and instructions provided as to how to use them with the units of astronomy.

In order to deal with a real universe using any cosmological equations, test particles that can be observed must be defined. These particles are not as small as individual stars or even galaxies. We do not know what scale size applies to these test particles. They must be large enough that their dynamics are controlled by the dynamics of space itself, that is by cosmology. So we need to know whether motions observed are 'cosmological' or 'non-cosmological'. This explains why nearby galaxies are not adequate as test particles. Many galaxies have their motions perturbed by the gravitational influence of other nearby galaxies. Even the motions of thousands of galaxies in rich clusters, such as the Virgo or Coma Clusters, can be controlled by the gravitational forces within the clusters themselves. The

Virgo Cluster of galaxies, which is the closest rich cluster to us, has an average recessional velocity of about 1100 km s^{-1} relative to our Galaxy. Yet, a few individual galaxies in the Virgo Cluster are actually approaching us, seeming to oppose the statement that the universe is expanding. The Coma Cluster has a median redshift of about 7000 km s^{-1}, but a velocity dispersion among the galaxies of the cluster of 900 km s^{-1} (Rood *et al.* 1972).

These samples illustrate that we must be careful in interpreting the motions of galaxies and in deconvolving cosmological and non-cosmological motions. Certainly, we know that the motions of galaxies within clusters do not reflect the behavior of the cosmological substratum of space. Yet, an overall cosmological expansion is unquestionably demonstrated using the median velocities of many clusters (Hoessel, Gunn & Thuan 1980, Kristian, Sandage & Westphal 1978). The clusters of galaxies seem to be self-contained, gravitationally bound nests that internally obey newtonian dynamics while acting as cosmological test particles relative to the rest of the universe. There are difficulties, however, in defining where one cluster stops and another begins; the universe is not constructed simply of isolated clusters immersed in otherwise empty space (Davis *et al.* 1982, Tully 1982).

There are galaxies, the 'field galaxies', which are not obviously members of any cluster. It might be expected that such galaxies would have small non-cosmological velocity components, and this seems to be the case. It has been claimed that these components are small indeed. Sandage & Tammann (1975*a*) state: 'An upper limit to the mean random motion of the field galaxies here can be put at about 50 km s^{-1} . . . The local velocity field is as regular, linear, isotropic, and quiet as it can be mapped with the present material.' This statement is inconsistent, however, with results that show our own Galaxy to be moving at several hundred km s^{-1} relative to the microwave background and reference frames provided by distant galaxies (Gorenstein & Smoot 1981, Aaronson *et al.* 1980, de Vaucouleurs & Peters 1981). It appears that the possibility of a non-cosmological velocity component of a few hundred km s^{-1} has to be allowed for any galaxy, whether a cluster member or not.

The fact that nearby field galaxies show significant cosmological motions was critically important to the beginnings of observational cosmology. It was these galaxies whose redshift measurements stimulated cosmological models based upon an expanding universe. By current standards, observations of these same galaxies would be very weak proof of such cosmologies. The 24 galaxies in the seminal paper on the subject (Hubble 1929) extend to redshifts of only 1100 km s^{-1}. The weak trend for redshift to increase with distance provided the foundation of observational cosmology. An analogous astronomer inhabiting the central regions of the Virgo Cluster would not have been able to discover the expanding universe because of the much larger, gravitationally induced velocity dispersion of the galaxies near him.

Despite the complications alluded to, observational cosmology today is firmly grounded on the presumption that galaxies have redshifts that are proportional to their distance, except for small components of non-cosmological motions which rarely exceed a few hundred km s^{-1}. It is safe, therefore, to consider galaxies with redshift velocities above several thousand km s^{-1} as having cosmological motions, to within acceptable uncertainty.

What about quasars? Can we show empirically that redshift is dependent only on distance for them as well? There is no single quasar for which it has been possible to measure distance, as can be done for nearby galaxies, using individual stars such as Cepheid variables. Such 'primary' distance indicators are at present useful only to distances of a few Mpc. Even for galaxies, most distance estimates arise from 'tertiary' indicators, based on angular diameters or total luminosities of entire galaxies (Sandage & Tammann 1975b, de Vaucouleurs 1979). The tertiary indicators are calibrated by comparison with closer galaxies whose distances can be measured more confidently. For quasars seen to be associated with galaxies, such tertiary indicators can be used. When this is done, quasars appear to obey the same redshift–distance relation as other galaxies (Chapter 4). If, as evidence summarized later indicates, quasars are events associated with galaxies, they must obey the same cosmological relations, so cosmologies derived from motions of galaxies can also be applied to quasars. Except for a handful of nearby Seyfert 1 galaxies, all quasars have sufficiently large redshifts that gravitationally induced motions of a few hundred km s^{-1} would be negligible in comparison. From this consideration, we consider that all quasars act as cosmological test particles. They cannot be used to determine the correct cosmology for the universe, because sufficiently precise distance indicators do not yet exist for quasars. But quasars must obey the rules defined by cosmologies derived from galaxies, so these cosmologies can be used to describe the relation between redshift and distance for quasars. This makes it possible to relate observed and intrinsic properties for quasars.

3.3   A newtonian cosmology

To illustrate the way in which these relations are produced, I first proceed to review the cosmological equations for a newtonian universe, which will have very close analogs in the correct relativistic equations. The only observables for most quasars are flux (or apparent magnitude) and redshift. Only a minority are close enough that an associated galactic disk can be resolved with present techniques. So flux and redshift are usually all that are available to transform to a distance or an absolute luminosity. Consequently, the target must be an equation relating redshift to distance, or relating flux and redshift to absolute luminosity. A newtonian cosmology provides the simplest illustration of how to achieve such equations (Rowan-Robinson 1977).

To approximate the universe with a newtonian view, consider a cloud of particles that is uniformly expanding. These are the test particles, and they are distributed within the cloud with uniform density. This accomplishes two necessary conditions: the motions of the particles are affected only by the overall ensemble of particles in the cloud, not by nearby density irregularities; and for an observer whose view within the cloud did not reach the edge, the cosmological principle would appear to be satisfied. For an observer in the cloud, the expansion of this cloud of particles is, therefore, analogous to the cosmological expansion of the universe. As that observer adopts the cosmological principle, he can apply any conclusions reached about the nature of the motions in his portion of the cloud to any other portion, seen or not. So it is only necessary to consider the motions of a small volume of the cloud, taken for convenience to be spherical. This approximation would be valid for the real universe if the sphere were large enough to even out density inhomogeneities for the test particles.

The departure from an approximation of the real universe comes with the desire to make simultaneous measurements of distances within the test sphere. Light travel times from particle to particle are ignored, which cannot be done for a volume in the real universe that is large enough to be approximated as homogeneous. At any time when observations are made, the test sphere has a certain size, which is characterized by radius r, which is $r(t)$ because the sphere is expanding along with the rest of its universe. It is conventional with any cosmology to normalize observations to the present and to denote observations made now with the subscript o. There will then be an r_o and a t_o. The concept of a cosmological expansion is that the entire universe enlarges uniformly; all test particles must keep the same relative configuration, with distances between all pairs increasing by the same factor. This is equivalent to a magnification of the entire ensemble. That is, any and all distances increase with the same scale factor. This dimensionless scale factor is normally denoted $R(t)$, so that applying it to any distance measured at time t_o would give that distance at time t. The r is then $r(t) = R(t)r(t_o) = R(t)r_o$, with the obvious normalization that $R(t_o) = 1$. Likewise, the distance $d(t)$ between any pair of particles is $d(t) = R(t)d_o$. This gives rise to the existence of an observed velocity–distance relation in this cosmology. An observer on any particle watching another particle recede with the general expansion of the system will see a redshift velocity $d'(t) = R'(t)d_o$, so that velocity $d'(t)$ will always be proportional to distance d_o for observations made at the same time t.

One of the properties of newtonian gravitation is the r^{-2} dependence of force, which results in the force at the surface of a sphere immersed in a homogeneous medium much larger than the sphere being an attractive force toward the center of the sphere and whose magnitude depends only on the matter within the sphere. This situation suits the idealized newtonian universe we are dealing with, so any particle of mass m at the edge of a sphere of radius r feels only a force F given by

$$F = -4\pi G m r^3 \rho / 3 r^2 = -4\pi G m \rho r / 3, \tag{3.1}$$

where ρ is the density of the matter within r. Because the sphere is expanding, $\rho = \rho_o / R^3$. This attractive force is the only force available to slow the expansion of this newtonian sphere. As a result, the radius r must relate to F by $F = mr''$. Assembling these defining equations gives the result that

$$R'' = -4\pi G \rho_o / 3 R^2, \tag{3.2}$$

recalling the definition of R. This is also a good time to point out another implicit subtlety: the assumption that G is not a function of time. This assumption is made for all cosmologies to be used below, and is a nearly unanimous choice in astronomy, but it is based only on the absence of any evidence for a time-varying G.

3.3.1   Open and closed universes in a newtonian cosmology

Equation 3.2 for R'' already provides an interesting result: that a newtonian universe cannot be static, but must be either expanding or contracting. Given these 50–50 odds, the idea of an expanding universe no longer seems extraordinary. The reason for this result is that a static universe, by definition, has $R' = 0$, which also yields that $R'' = 0$. This can only satisfy equation 3.2 for R'' if $\rho_o = 0$. So, as long as there is any matter at all in a newtonian universe, it must have a non-zero R'. The detailed nature of $R(t)$ depends on the initial conditions; that is, the balance between the impulse that set the system expanding initially and the gravitational attraction among the particles in the system that is tending to stop the expansion. The next step, therefore, is to examine the integrals of R'' to consider other forms of the equation of motion for this system. The first integral is easily done after multiplying through by $2R'$, yielding

$$2R'R'' = -8\pi G \rho_o R^{-2} R' / 3,$$

the integral of this being

$$R'^2 = (8\pi G\rho_0 R^{-1}/3) - k. \tag{3.3}$$

The constant of integration k controls the destiny of this cosmological model, and there are only three possibilities: k is zero, k is positive, or k is negative. These alternatives determine whether this universe is open or closed, unbound or bound, capable of infinite expansion or not. The most important alternative is $k = 0$, because it will be quickly shown that this provides the dividing case between an open and closed universe. If $k = 0$, $R'^2 R = 8\pi G\rho_0/3$, and the expression for R which satisfies this equation is $R = (6\pi G\rho_0)^{1/3} t^{2/3}$. Physically, this states that R', the expansion velocity, cannot go to zero unless R becomes infinite, for any finite ρ_0. The universe, therefore, is expanding just at the escape velocity when $k = 0$; the expansion velocity progressively slows but approaches zero only asymptotically. If the expansion velocity is greater than the escape velocity (negative k), the expansion of the universe will never cease, and the universe is said to be open. If the expansion velocity is below the escape velocity (positive k), the expansion will eventually cease, reverse to a contraction, and the universe will fall back into itself. In the latter case, the universe is said to be closed. The characteristics of a real newtonian universe would be on one side or the other of the $k = 0$ borderline case, so it is convenient to describe the observational parameters of the universe in context of this case. (In the relativistic universes to be discussed later, this case is the Einstein–de Sitter universe, for which space is 'flat'. For convenience, therefore, this case is often referred to as a flat universe.)

Clearly, the expansion velocity R' changes with time, so the R' observed today did not apply throughout the entire time the universe has been expanding until the present. Because the matter in the universe must have been slowing this expansion, the universe was expanding faster in the past, though by how much is not known until the value of k is known. In any case, we can certainly derive an upper limit to the age of the universe by assuming $R'(t) = R'(t_0)$. This limiting age is called the Hubble time $= T_0$, defined as $T_0 = R(t_0)/R'(t_0)$. The inverse of T_0 is defined as the Hubble constant $= H_0 = T_0^{-1}$. With the introduction of these parameters, it is possible to make contact with observational astronomy.

Recall that distance between any two test particles in this model universe is given by $d = R(t)d_0$. Observed distances d_0 to other test particles in the real universe are conventionally measured in megaparsecs (1 Mpc $= 3.08 \times 10^{24}$ cm). The mutual recessional velocity between any two particles, or the redshift velocity of one particle as observed from the other, is $R'(t)d_0$. This velocity is conventionally measured in km s^{-1}. The Hubble constant can be determined, therefore, with an observation of distance and velocity for only one test particle, because $H_o = R'(t_o)/R(t_o) = d'(t_o)/d(t_o)$. H_o is quoted in units of km s^{-1} Mpc^{-1}, the inverse of which has units of seconds if Mpc are converted to km. Shortly, actual values for H_o and T_o will be discussed. But the observed T_o is only an upper limit to the time of expansion; how does this relate to the actual time for which the expansion has been underway?

For the $k = 0$ case under consideration, the answer is easy. Because the equation for $R(t)$ goes as $t^{2/3}$, the $T = R/R'$ is $T = 3t/2$. This means that the actual time t for which the expansion has been underway ($R = 0$ when $t = 0$) is only two-thirds of the measured Hubble time T. This provides a critical benchmark for scaling the real age of the universe to the observed Hubble time, and consequences of this are considered below. There remain two alternatives for k to be addressed first, however.

As previously stated, a positive k corresponds to a bound, or closed, universe. This is because R' can become zero, in equation 3.3, when R has a value R_{max} such that $8\pi G\rho_o/3R = k$. The maximum size the universe will reach, R_{max}, could be found, and it depends on ρ_o and the numerical value of k, which sets the scale of the initial expansion impulse. Equation 3.3 also shows that positive k requires a larger ρ_o to achieve the same combination of R and R' as in the $k = 0$ case. From equation 3.2, this requires a more negative R''; i.e. the universe must have decelerated such that $t < 2T_o/3$. The actual age is $< 2H_o^{-1}/3$.

Finally, for negative k, the equation 3.3 shows that R' reaches a non-zero positive value as R approaches infinity, so the expansion of the universe never ceases, and the universe is open. If k is sufficiently negative, it dominates the right-hand side of equation 3.3 and t approaches T, since R' changes little with time. This is the only case in which the actual expansion age of the universe can be indistinguishable from the Hubble time.

3.3.2   Observational constraints on open vs. closed

With these alternatives for the expansion age in mind, it is now possible to compare with existing observations. Those of the Hubble constant are the most critical and still the subject of major controversy. Various approaches to measuring H_o have been described recently in extensive detail. Major and reasonably independent efforts are summarized by Aaronson et al. (1980) giving $H_o = 95 \pm 5$ km s^{-1} Mpc^{-1}; by de Vaucouleurs & Bollinger (1979) giving $H_o = 100 \pm 10$; and by Sandage & Tammann (1982) giving $H_o = 50 \pm 7$. There is not space in this volume for a critical review of the H_o controversy and the reasons for the differences found. Astronomers are able to get the same answer when observing the same thing. The difficulty with determining H_o is that distance estimators are required for galaxies, and there are many things that can potentially be used as such estimators. Depending upon the weight assigned to the various alternatives, different galactic distances might be found. Furthermore, determining H_o requires making corrections for any non-cosmological motion of our own Galaxy, since that is our observing platform, and there are conflicting conclusions about that motion (Section 3.2).

For the moment, the most important statement is that good evidence has been presented by more than one group that $H_o > 90$ km s^{-1} Mpc^{-1}. Until the observations yielding this result are proven wrong, this is a matter of major concern for existing cosmologies. As can be seen from the foregoing discussion of expansion age, the maximum allowable age for the universe is H_o^{-1}, or $t < 11 \times 10^9$ y, for $H_o > 90$. This age is less by several billion years than the empirical ages claimed by persons who model the evolution of stars in the oldest observed star clusters (Carney 1980, Sandage 1982). If the universe is flat or closed, this age discrepancy is even more dramatic as the maximum allowable expansion age decreases to $t < 2H_o/3$, or $t < 7.4 \times 10^9$ y for $H_o > 90$.

This demonstrates the first of the two major puzzles mentioned in the introduction to this chapter that must be resolved before we can be entirely comfortable with Friedmann cosmologies. An astute reader will perceive that the age discrepancy would disappear, even if $H_o > 90$, if there were a way of accelerating the universe, so that t could exceed T. Within a newtonian cosmology, this is impossible because there is no long range repulsive force to accomplish this acceleration. Nor does such a force exist in relativistic cosmologies, but these can accommodate space whose expansion accelerates. A term in the relativistic equations – the 'cosmological constant' – is available for this purpose, although utilization of it violates existing laws of physics. So this term is conventionally taken to be zero, and that convention is likely to stand until there is no longer room for escape by reducing H_o, globular cluster ages, or both in adequate combination to avoid the need for an accelerating universe.

3.4   Deceleration parameter q and density parameter Ω

The second major puzzle of cosmology that leaves nagging doubts as to the fundamental correctness of our models is the inability to decide if the universe is open or closed. Why should nature have placed the universe so close to the boundary between these alternatives? That this is indeed the case can be seen by examining the parameters of a simple newtonian universe, but it is convenient to introduce the notation used for this examination within relativistic cosmologies. So far, the behavior of a newtonian universe has been characterized by the parameter k, but observational cosmology and relativistic models commonly use the deceleration parameter q. This is defined as $q = -RR''/R'^2$. Since the boundary case between an open and closed universe has $k = 0$, for which $R = (6\pi G\rho_0)^{1/3}t^{2/3}$, $R' = (2/3)(6\pi G\rho_0)^{1/3}t^{-1/3}$, and $R'' = (-2/9)(6\pi G\rho_0)^{1/3}t^{-4/3}$, the value of q for this case is $q = 0.5$. For a closed universe with positive k, R'' was more negative than for $k = 0$, which would make q more positive, because R and R' must be positive. Similarly, for an open universe with negative k, R'' is more positive than in the $k = 0$ case, making q more negative. All in all, this means that an expanding universe with $q > 0.5$ is closed, and one with $q < 0.5$ is open. As long as no acceleration is allowed (meaning R'' always negative), note that q can never be negative. Within cosmologies based only upon the attractive force of gravity, therefore, any open universe must have $0 < q < 0.5$, which appears quite restrictive. Attempts – heroic efforts, really – have been made to measure q_0 directly by observing sufficiently distant in the universe to see how R' has changed as the universe expanded, and so to determine if the deceleration is as rapid as expected in a closed universe (Kristian *et al.* 1978, Hoessel *et al.* 1980). These attempts have, so far, been frustrated by the inability to define a 'standard candle', a test particle whose distance can be reliably measured at large redshift. The observations have been made by observing the first-ranked (brightest) galaxies in rich clusters, as these would be expected to be similar galaxies. The difficulties that arise are not in finding or measuring such galaxies; the problem is in not knowing how the luminosity of the first-ranked galaxy evolves with redshift.

Alternatively, it should be possible to decide what q_0 has to be by measuring the density of the local universe, because the density controls R''. Appealing once more to the cosmological principle, it is necessary to decide only if the local sphere of the universe that can be analyzed with newtonian equations has sufficient density to stop its own expansion. If it does, the entire universe must stop as well, because no one part can behave differently from another. It was determined for a newtonian universe that $R'^2 = 8\pi G\rho_0/3R$ when $k = 0$ (or $q = 0.5$). Were ρ_0 to satisfy this equation, it would mean that the density of the universe is just sufficient to stop its own expansion. That particular value of the density is defined as the critical density, ρ_c. ρ_0 is not a function of time, being defined as the density of the universe today; ρ_c is a function of time, because the density required to stop the expansion will decrease as R increases. Or, $\rho_c = \rho_0/R^3$, and $R'^2 = (8/3)\pi G\rho_0 R^{-1}$. The critical density for the universe, that density that at any time would be just sufficient to stop the expansion as R approaches infinity, is then $\rho_c = 3H^2/8\pi G$, because $H = R'/R$.

At any epoch in the universe, it is only necessary to compare the actual density at that epoch with the critical density to determine whether the universe is open (actual density less than critical), flat, or closed (actual density more than critical). This density ratio makes a convenient cosmological parameter, defined as Ω, with Ω = actual density/critical density. In the local universe, today, $\Omega_0 = \rho_0/\rho_c = \rho_0(3H_0^2/8\pi G)^{-1}$. If we are able to measure H_0 and ρ_0, we can decide the destiny of the universe. This is consistent with previous conditions on q, because from equation 3.2, $4\pi G\rho = -3R''/R$ so $\Omega = \rho/\rho_c = -2R''R/R'^2 = 2q$. As we already knew that $q = 0.5$ is the dividing case, this is the same as $\Omega = 1$. For $q > 0.5$, or $\Omega > 1$, the universe is closed; for $q < 0.5$, or $\Omega < 1$, it is open; and q cannot be less than zero as that would give a negative Ω, implying the actual density of the universe to be negative. Inserting the range of values for H_0, the critical density of the local universe today is $5 \times 10^{-30} < \rho_c < 2 \times 10^{-29}$, for $50 < H_0 < 100$ km s^{-1} Mpc^{-1} and ρ_c in g cm^{-3}.

Conceptually, determining Ω_0 is therefore a simple matter of measuring the average density of the local universe. Observationally, this is a major challenge. It is necessary to determine ρ_0 over a sufficiently large volume that local inhomogeneities are smoothed out. We are faced once again with deciding how large a volume of the universe must be observed to be a representative sample, so all the uncertainties mentioned earlier regarding galaxy clustering arise again. Furthermore, the bulk of the mass in the universe is invisible, if the mass determined dynamically is compared with that which can be seen in galaxies (Faber & Gallagher 1979). It is necessary to measure the mass density of the universe by measuring the non-cosmological motions which this mass causes for galaxies. As might be expected, this is not easy. Comprehensive analyses of the motions of field galaxies yield $\Omega_0 \approx 0.2$ (Davis & Peebles 1982).

This measured Ω_0 refers not only to the visible stars in galaxies, but includes any dark matter as long as the latter is distributed similarly to the galaxies; were the dark matter distributed more evenly than galaxies, Ω_0 would be larger. Such mysterious entities as massive haloes of galaxies are included. In fact, an Ω_0 applying only to visible matter is of order 0.01. The puzzle of the 'missing matter' is, consequently, already included when Ω_0 of 0.1 or more are found. Any matter, including massive neutrinos or other non-baryon material, is included unless this matter is distributed more homogeneously than the visible galaxies. The galaxies serve as gravitationally influenced tracers of whatever density inhomogeneities exist; the galaxies should trace the distribution of matter in the universe no matter what is the nature of this matter.

So the results for Ω_0 seem to give the answer: the universe is open, so open that it almost acts dynamically as if it were empty of matter. From this consideration of Ω_0, $q_0 = 0.1$. Recall that having q_0 close to zero is desirable because it is then easier to avoid the time scale difficulties discussed previously. Adopting a small q_0 is comforting in this regard.

We might now safely proceed by applying cosmological equations for an open universe, were it not for the curiosity of having a universe so close to the critical value of unity for Ω_0. Given that Ω_0 can range over all possible values, from extremely large to extremely small, the thought has struck many cosmologists that it is exceedingly curious for it to be so close to unity. Why is the universe so nearly flat? One growing school of cosmology argues that the observations must be slightly wrong and the universe really has Ω_0 precisely unity. An implication is that observations of H_0 and/or cluster ages have to be wrong in sufficient combination to overcome the age inconsistency, which is severe if $q_0 = 0.5$. Theoretically, having a flat universe would be so meaningful that an explanation for it could be sought. Seeking such an explanation was one motivation for the 'inflationary' cosmologies (Guth 1981). This philosophical *tour de force* may prove to be correct. It seems reasonable to give it sufficient credence that the cosmological equations for a flat universe are provided. In the equations to be applied to quasars, therefore, alternates will be given for $q_0 = 0.5$ and $q_0 = 0.1$, with the implication that these bracket the actual nature of the universe, to within the understanding of cosmology at present.

3.5   Relativistic cosmologies

The simple derivations presented previously for the equations of motion of a newtonian universe were a guide to the procedures and assumptions used to construct cosmologically useful equations. Analogous procedures, though much more complex mathematically, have been used to construct the field equations of relativistic cosmologies (Weinberg 1972, Peebles 1971). Unlike a newtonian universe, such equations incorporate a space curvature as the constant of integration. (This provides the mathematical necessity for the cosmological principle within relativistic cosmologies, because the same curvature must exist throughout the universe.) In flat space ($\Omega_0 = 1$), this curvature term is zero, which is clearly a simpler case; this is the Einstein–de Sitter universe. The simplest expression of a general metric allowing use of any curvature constant is the Robertson–Walker metric. Use of this metric with the assumption that the cosmological constant is zero (no intrinsic acceleration of the universe) produces the class of Friedmann universes, one of which will be utilized below for $q_0 = 0.1$.

To reach the equations that will be applied to observations of quasars, it is necessary to begin by considering the fundamental aspects of the propagation of light through the universe. Within euclidean space, and commonplace experience, the flux observed relates to the intrinsic luminosity radiated by

$$f = L/4\pi d_l^2, \tag{3.4}$$

which incorporates a fundamental definition: d_l is the distance an object appears to have if it is assumed that the inverse square law relating luminosity and flux is precisely valid. This is not the case in any curved space, so this 'luminosity distance' is not the only concept of distance. More fundamental in relativistic universes is the 'proper distance' defined by light travel time along a null geodesic. If a light signal is emitted at one event, such as a galaxy at a particular time, and received by another, such as an astronomer at another time, these events are connected only by a light signal. These are the circumstances for everything we observe in the universe, as opposed to connections via actual travel by matter through the universe. Because all observers, regardless of the velocity or acceleration of their reference frames, must measure the same speed of light, the connection of events by a light signal travelling along a null geodesic defines the most basic concept of distance that can be applied to the universe. The objective is to relate the luminosity emitted at time t_e to the flux observed at time t_o, using light which has travelled a null geodesic from source to observer.

3.5.1   Fluxes and redshifts

To relate intrinsic and observed parameters, we use ν_e and λ_e for the emitted frequency or wavelength, and ν_o, λ_o for the observed frequency or wavelength. These relate by the redshift parameter z, defined such that

$$\nu_e/\nu_o = \lambda_o/\lambda_e = 1 + z. \tag{3.5}$$

For a null geodesic, $c\,dt/R(t)$ is a constant independent of time. So, this ratio is the same at any time, or $c\,dt_e/R(t_e) = c\,dt_o/R(t_o)$ giving $dt_e/dt_o = R(t_e)/R(t_o)$. As frequency is in (units of time)$^{-1}$, $\nu = 1/dt$, so $\nu_e/\nu_o = dt_o/dt_e = R(t_o)/R(t_e)$. Having defined $\nu_e/\nu_o = 1 + z$, it must be that $R(t_o)/R(t_e) = 1 + z$.

This shows what a fundamental parameter z really is, allowing a determination of the change in the scale factor of the universe between the time of emission and observation of a light signal. This result also gives the first key modification to the concept of distance. It might be assumed that the proper distance along a null geodesic must reduce to the luminosity distance d_l for sufficiently small values of z, analogous to the way in which relativistic and newtonian concepts must reduce to the same equations in the limit of very small velocities of relative motion. That is, for sufficiently small distances in the universe, all space must seem euclidean. In this limit, we can expect $f_o = L_e/4\pi d^2$, with d the proper distance. Paying attention to units for the first time, f_o is measured in erg cm^{-2} s^{-1} and L_e in erg s^{-1}. Both quantities are, so far, bolometric properties because these units imply that we are capable of measuring the total flux or luminosity over all frequencies.

Something basic changes as z increases, because the 'second' in the observer's frame is not the same as the 'second' in the emitter's frame, these being related by $dt_o = (1 + z)dt_e$. A unit interval of time is longer for the observer by a factor of $1 + z$. This effect then supplies a factor of $1 + z$ that must be applied in transforming from observed fluxes to emitted luminosities. Another factor of $1 + z$ is also necessary. The change in unit time between emitter and observer slows the photon arrival rate at the observer, but each photon is redshifted in frequency by $1 + z$, so each photon carries less energy by a factor of $1 + z$ than when emitted. As the flux and luminosity are being related in energy units, the net effect of these two conceptual corrections is to make

$$f_o = L_e/4\pi d^2(1 + z)^2. \tag{3.6}$$

With this result, the luminosity distance has become $d(1 + z)$. The next step is to determine proper distance d as a function of z.

Determining $d(z)$ requires obtaining the integral of the line element along the null geodesic from observer to emitter. It would be anticipated that, at its simplest, the result would have to depend on space curvature, which can be described by the value of q, and on the relation between units for measuring redshift and those for measuring distance, which is described by H. With these parameters, the solution for proper distance (Sandage 1975, Weinberg 1972) is given by:

$$d = c\{q_o z + [q_o - 1][(1 + 2q_o z)^{0.5} - 1]\}/q_o^2 H_o (1 + z)$$
$$\text{for } q_o > 0, \tag{3.7}$$

simplifying to

$$d = cz\left(1 + \frac{z}{2}\right)\Big/ H_o(1 + z) \text{ for } q_o = 0.$$

Recalling that we intend to bracket what we hope is the true condition in the universe by using alternate q_o of 0.5 and 0.1, we can now produce the first set of working cosmological equations for use with quasars:

$$f_o = L_e/4\pi c^2 H_o^{-2}\{10z - 90[(1 + 0.2z)^{0.5} - 1]\}^2, \quad q_o = 0.1; \tag{3.8a}$$

$$f_o = L_e/16\pi c^2 H_o^{-2}[(1 + z) - (1 + z)^{0.5}]^2, \quad q_o = 0.5. \tag{3.8b}$$

To use conventional astronomical units, H_o is quoted in km s^{-1} Mpc^{-1}, but f_o is in erg cm^{-2} s^{-1} and L_e is in erg s^{-1}, so inserting numerical values with these units leads to:

$$f_o = 9.32 \times 10^{-62} \, L_e H_o^2 / \{10z - 90[(1 + 0.2z)^{0.5} - 1]\}^2,$$
$$q_o = 0.1; \qquad\qquad\qquad\qquad\qquad\qquad\qquad (3.9a)$$
$$f_o = 2.33 \times 10^{-62} \, L_e H_o^2 / [(1 + z) - (1 + z)^{0.5}]^2,$$
$$q_o = 0.5. \qquad\qquad\qquad\qquad\qquad\qquad\qquad (3.9b)$$

These relations among f_o and L_e apply only to bolometric quantities, which in practice are highly idealized because of the difficulty of measuring the entire luminosity of a quasar over all frequencies. It is often desirable, however, to use the luminosity of a single emission line as a fundamental parameter. When this is done by measuring the total flux in an emission line, regardless of the line profile or width, equations 3.9a and b apply. This is the only circumstance for which these particular equations prove useful.

It is more common to measure fluxes and luminosities in the continuous spectra of quasars. To provide cosmological equations for this circumstance, the definitions used are:

f_{vo} = flux observed in units of erg cm^{-2} s^{-1} Hz^{-1},

L_{ve} = luminosity emitted in erg s^{-1} Hz^{-1},

$f_{\lambda o}$ = flux observed in erg cm^{-2} s^{-1} Å^{-1},

$L_{\lambda e}$ = luminosity emitted in erg s^{-1} Å^{-1}.

Recall that for bolometric luminosities, $L_e = 4\pi d_1^2 f_o$ with the relations for d_1 depending on q_o. But we are now considering flux measured at a wavelength or frequency in the observer's reference frame, and luminosity arising at a different wavelength or frequency in the emitter's frame. If observing f_{vo}, the frequency interval $dv_e = (1 + z)dv_o$, because $v_e = v_o(1 + z)$. This means that a frequency interval dv_o in the observer's frame only includes $(1 + z)^{-1}$ as much spectrum as was emitted in the quasar's frame. Imagine an object that emitted L_{ve} over a bandpass of width dv_e and had zero luminosity at all other frequencies. Then, the bolometric luminosity emitted is $L_{ve}dv_e$ and the bolometric flux observed is $f_{vo}dv_o$, with $dv_e = (1 + z)dv_o$. So, over this bandpass $L_{ve}dv_e = 4\pi d_1^2 f_{vo}dv_o$ so

$$L_{ve} = 4\pi d_1^2 f_{vo}dv_o/dv_e = 4\pi d_1^2 f_{vo}(1 + z)^{-1} = 4\pi d^2 f_{vo}\,(1 + z).$$

For observations per unit wavelength interval $d\lambda_e = (1 + z)^{-1}d\lambda_o$ and $L_{\lambda e}d\lambda_e = 4\pi d_1^2 f_{\lambda o}d\lambda_o$, giving

$$L_{\lambda e} = 4\pi d_1^2 f_{\lambda o}\,(1 + z) = 4\pi d^2 f_{\lambda o}(1 + z)^3.$$

Inserting the proper forms of d for $q_o = 0.5$ and 0.1,

$$f_{\nu o} = 9.32 \times 10^{-62} \, L_{\nu e} H_o^2 (1 + z)/\{10z - 90[(1 + 0.2z)^{0.5} - 1]\}^2,$$
$$(3.10a)$$

$$f_{\lambda o} = 9.32 \times 10^{-62} \, L_{\lambda e} H_o^2/(1 + z)\{10z - 90[(1 + 0.2z)^{0.5} - 1]\}^2,$$
$$q_o = 0.1; \qquad\qquad (3.10b)$$

and

$$f_{\nu o} = 2.33 \times 10^{-62} \, L_{\nu e} H_o^2 (1 + z)/[(1 + z) - (1 + z)^{0.5}]^2,$$
$$(3.11a)$$

$$f_{\lambda o} = 2.33 \times 10^{-62} \, L_{\lambda e} H_o^2/(1 + z)[(1 + z) - (1 + z)^{0.5}]^2,$$
$$q_o = 0.5. \qquad\qquad (3.11b)$$

The numerical factor accommodates the transformation of units when H_o is in km s^{-1} Mpc^{-1}, and fluxes and luminosities have the units defined above.

Another parameter often used in spectroscopy is the equivalent width. This is a measure of the strength of an emission or absorption line, compared with that of the adjacent continuum. Equivalent width W is defined as the wavelength interval of the continuum that contains the same flux as the total flux in the line considered. That is,

$$W_o = f_o(\text{line})/f_{\lambda o},$$

where $f_{\lambda o}$ is the continuum flux interpolated at the position of the line center. This is the observed equivalent width, but the equivalent width in the quasar frame would be

$$W_e = L_e(\text{line})/L_{\lambda e}.$$

So the observed and emitted equivalent widths can be related by

$$W_o = (L_e/4\pi d_1^2)/[L_{\lambda e}/4\pi d_1^2(1 + z)] = W_e(1 + z). \qquad (3.12)$$

This effect has the consequence of making emission lines from quasars appear more conspicuous as the redshift increases. As many surveys for quasars depend on recognizing emission lines in low resolution spectra (Chapter 2), this effect is especially helpful when searching for quasars at high redshifts that are faint.

3.5.2   K-corrections

The transformations discussed so far make it possible to go from observations to parameters which describe intrinsic properties of the quasar. Even then, a further step is needed to relate the same intrinsic properties among all quasars, because the observer sees a different part of the intrinsic spectrum, depending on the redshift of the quasar observed. To compare an object at one redshift with those at others, it is necessary to normalize to some standard wavelength or frequency within the object's spectrum. This is the source of the concept called the K-correction, introduced originally to deal with a similar problem in the observation of distant galaxies (Sandage 1975). The K-correction as now formally defined for galaxies includes both the correction for the luminosity change as a function of λ_e and the change in bandwidth $d\lambda_e$ to $d\lambda_o$. As the bandwidth change is already incorporated in equations 3.10 and 3.11, it is only necessary here to describe the correction that arises from the changing $L_{\nu e}$ or $L_{\lambda e}$ within a continuous spectrum. Once an intrinsic luminosity has been observed at any place in the continuous spectrum of an object, that result can be used to determine the luminosity of any other part of the spectrum if the intrinsic form of the spectrum is known.

For quasars, the K-correction is straightforward to deal with as long as continuous spectra of quasars are representable by power laws; that is, the intrinsic luminosities are $L_{\nu e} = K\nu_e^\alpha$, where K is a constant for a given quasar and α is the power law index. To convert into $L_{\lambda e}$ units, it must be that $L_{\nu e}d\nu_e = -L_{\lambda e}d\lambda_e$ for $d\nu_e$ and $d\lambda_e$ corresponding to the same interval. (The negative sign arises because increasing λ corresponds to decreasing ν.) Since, in general, $\nu = c\lambda^{-1}$, $d\nu/d\lambda = -c\lambda^{-2}$ so

$$L_{\lambda e} = L_{\nu e}c\lambda_e^{-2} = K\nu_e^\alpha c\lambda_e^{-2} \propto \lambda_e^{-(2+\alpha)}.$$

This means that any power law spectrum for which

$$L_{\nu e} \propto \nu_e^\alpha \text{ must also have } L_{\lambda e} \propto \lambda_e^{-(2+\alpha)}.$$

This index is not the same for all quasars or even within different parts of a single quasar spectrum (the X-ray may differ from the optical and infrared, which may differ from the radio; see Chapter 7). As long as a single value α applies over the range between the emitted λ_e and $(1 + z)$ times emitted λ_e, relations can be derived that allow $L_{\nu e}$ or $L_{\lambda e}$ to be determined for any ν_e or λ_e. Equations 3.10 and 3.11 could then be modified to determine any L_ν or L_λ by substituting $L_{\nu e} = L_\nu(\nu/\nu_e)^{-\alpha}$ or $L_{\lambda e} = L_\lambda(\lambda/\lambda_e)^{2+\alpha}$.

For convenience, the following relations illustrate the way to determine $L_{\nu o}$ or $L_{\lambda o}$, which give the intrinsic luminosity of the quasar at the *same* frequency or wavelength at which observations are made. This is especially convenient for comparing objects with negligible redshift with those at high redshift. For this case, $L_{\nu e} = L_{\nu o}(\nu_o/\nu_e)^{-\alpha} = L_{\nu o}(1 + z)^\alpha$, and $L_{\lambda e} = L_{\lambda o}(1 + z)^{2+\alpha}$. So,

$$f_{\nu o} = 9.32 \times 10^{-62} L_{\nu o}(1+z)^{1+\alpha}H_o^2/\{10z - 90[(1 + 0.2z)^{0.5} - 1]\}^2,$$
$$\tag{3.13a}$$

$$f_{\lambda o} = 9.32 \times 10^{-62} L_{\lambda o}(1+z)^{1+\alpha}H_o^2/\{10z - 90[(1 + 0.2z)^{0.5} - 1]\}^2,$$
$$q_o = 0.1; \tag{3.13b}$$

and

$$f_{\nu o} = 2.33 \times 10^{-62} L_{\nu o}(1 + z)^{1+\alpha} H_o^2 /[(1 + z) - (1 + z)^{0.5}]^2,$$
(3.14a)

$$f_{\lambda o} = 2.33 \times 10^{-62} L_{\lambda o}(1 + z)^{1+\alpha} H_o^2 /[(1 + z) - (1 + z)^{0.5}]^2, q_o = 0.5.$$
(3.14b)

In X-ray, ultraviolet, infrared and radio astronomy of quasars, flux and luminosity units are conventionally used. Optical astronomy of quasars, however, remains largely bound by the conventions of the past, which dictate use of magnitudes. Magnitudes are normally measured through filters that have a bandpass defined by their width in wavelength units. For example, the B magnitude arises from the flux that enters a filter about 1000 Å wide centered at about 4400 Å. To transform between magnitudes and fluxes, the effective wavelength of the magnitude system used has to be defined so a monochromatic flux can be described at a particular location in the spectrum. This transformation is further complicated by the arbitrary definition of the magnitude system, which sets zero magnitude at all wavelengths to correspond closely to the flux from α Lyrae (Vega). This star has different flux at different λ, so the relation between magnitude and flux depends upon the effective wavelength. Many relevant quasar observations are of B magnitudes, assumed here to correspond to 4400 Å effective wavelength, at which magnitudes and fluxes relate by (Hayes & Latham 1975):

$$B = -2.5 \log f_{\nu o}(4400 \text{ Å}) - 48.40$$
(3.15a)

or

$$B = -2.5 \log f_{\lambda o}(4400 \text{ Å}) - 20.42$$
(3.15b)

Magnitude units can be used exclusively, utilizing as a measure of intrinsic luminosity the definition of absolute magnitude M. M is defined as the apparent magnitude an object would have were it at a luminosity distance of 10 pc. This definition results in

$$B - M = -2.5 \log f + 2.5 \log \{L/4\pi[10\text{ pc} \times (3.08 \times 10^{18}\text{ cm pc}^{-1})]^2\}$$
$$= 2.5 \log (L/f) - 100.19.$$

Using equations 3.13 and 3.14,

$$B - M = 5 \log \{10z - 90[(1 + 0.2z)^{0.5} - 1]\} - 2.5(1 + \alpha) \log (1 + z)$$
$$- 5 \log H_o + 52.39, q_o = 0.1;$$
(3.16a)

and

$$B - M = 5 \log [(1 + z) - (1 + z)^{0.5}] - 2.5(1 + \alpha) \log (1 + z)$$
$$- 5 \log H_o + 53.89, q_o = 0.5.$$
(3.16b)

These last two equations are the ones to use if we wish to relate quasars observed at a given apparent B magnitude having differing redshifts, in order to determine their absolute magnitudes all at the same emitted wavelength. These are useful equations for interpreting studies of quasar evolution (Chapter 6).

3.5.3 Angular diameter and surface brightness

Although quasars are, by definition, spatially unresolved objects, they appear to be the nuclei of galaxies (Chapter 4). Often, they are found in clusters with other galaxies. It is sometimes desirable to determine the diameters of these associated galaxies, or the separations between quasars and other objects that may be physically associated with them. Similar needs apply for the extended radio sources associated with quasars, and the extended X-ray emission usually found in clusters of galaxies. For all these objectives, it is necessary to relate angular to linear measurements for various cosmologies. As observations press to fainter limits, and angular resolution capabilities increase, such measurements will become of increasing utility. How, then, do angular diameters depend on the cosmological parameters?

Consider a fixed distance D between, for example, two galaxies in a cluster. If this is a physically defined length, it always represents the same fraction of the scale size of the universe regardless of the epoch (redshift) at which it is measured. Such a measure is what is meant if we speak of a proper length, such as a kpc 'then' compared with a kpc 'now'.

Another way to state the concept is that proper lengths are those that would describe physical sizes if the universe were not expanding. The same applies to proper distance. In fact, the equations for proper length or proper distance are valid, and used in Chapter 4, for non-expanding cosmologies. The space curvature which governs the form of these equations is independent of whether or not the universe is expanding. So, if it is desired to describe the separations or diameters of objects, regardless of redshift in an expanding universe, a length that scales with R is necessary. From equation 3.5, therefore, the distance D observed from the local universe, D_o, would be $D_o = (1 + z)D$. As a result, the observed angular size in radians is D_o/d, where d is the proper distance. Looking to equation 3.7 for d, and converting from radians to arcseconds, gives angular size

$$\theta'' = 6.88 \times 10^{-4} H_o D (1 + z)^2 / \{10z - 90[(1 + 0.2z)^{0.5} - 1]\},$$
$$q_o = 0.1; \qquad (3.17a)$$
$$\theta'' = 3.44 \times 10^{-4} H_o D (1 + z)^2 / [(1 + z) - (1 + z)^{0.5}],$$
$$q_o = 0.5. \qquad (3.17b)$$

Both equations use kpc for units of D.

Equations 3.17*a* and *b* show a curious behavior not expected from euclidean geometry. The apparent angular size of length D does not significantly decrease with redshift at high z. If it becomes possible to measure, for example, the sizes of clusters of galaxies that contain quasars, this cosmological effect may become noticeable. To illustrate this, Table 3.1 lists the angular size corresponding to a length of 100 kpc (equivalent to several galaxy diameters or an approximate cluster radius) for $H_0 = 75$ km s^{-1} Mpc^{-1} and q_0 of 0.1 and 0.5. Note that such a size remains optically resolvable at all z, assuming the objects defining D remain visible. Whether this latter circumstance is true is an important question, discussed in Chapter 4.

As long as D refers to a distance between objects, there are no problems with the concept of what it means. It is a metric length. If, however, D is the diameter of a galaxy, it must be defined with care. A metric diameter corresponds to a diameter containing a defined fraction of the total luminosity of the galaxy, being a true physical quantity. Because galaxies have such fuzzy edges, the measurement of a metric diameter is difficult, requiring a luminosity profile across the galaxy. Instead, galaxy diameters are conventionally measured as isophotal diameters, meaning the diameter between locations where the apparent surface brightness of the galaxy has a particular value. For any resolved object – an optical galaxy, a radio lobe, hot gas in a cluster seen in X-rays – whose size measurement depends on its surface brightness, an isophotal rather than a metric diameter is measured. This is

Table 3.1. *Angular size for proper length of 100 kpc and* $H_0 = 75$

Redshift z	θ'' if $q_0 = 0.1$	θ'' if $q_0 = 0.5$
0.01	524	525
0.1	60	61
0.2	34	36
0.5	19	21
0.8	16	18
1.0	15	17.6
1.5	13.5	17.5
2.0	13.2	18.3
2.5	13.2	19.4
3.0	13.4	21
4.0	13.9	23
5.0	14.6	26

convenient and necessary observationally, because most astronomical measurements of resolved objects (the only exceptions being the Sun, Moon, and planets) are limited by the contrast between the surface brightness of the object and that of the background noise from sky or detector. Unfortunately, surface brightnesses and the isophotal diameters measured thereby are especially sensitive to cosmological effects. At high z, isophotal diameters have a complex dependence on several parameters, which are described in Chapter 4.

3.6 Look-back times

In all of the preceding discussions of luminosities, K-corrections, angular diameters, or any other properties of quasars, working relations have been described as a function of z. In general, the 'distance' to an object has been a quantity that depends on whether luminosity distance or proper distance is the preferred interpretation of distance. It is certainly useful and important, especially when considering quasar evolution, to consider the light travel time from an emitter to us, the observers. So, thinking of the distance to a quasar in terms of light years proves to be a fundamental concept; this is the idea of the 'look-back time'. This describes how long ago existed the epoch that is observed at redshift z. It is derived by finding the difference in the age of the universe between the time of light emission and the time of light observation.

From Section 3.5.1, the scale factors of the universe must have the ratio $R(t_o)/R(t_e) = 1 + z$. For the case of $q_o = 0.5$, it was shown in Section 3.3.1 that $R \propto t^{2/3}$. So $R(t_o)/R(t_e) = t_o^{2/3}/t_e^{2/3}$. The look-back time, commonly denoted τ, is $\tau = t_o - t_e$. Because we have for $q_o = 0.5$ that $t_o/t_e = (1 + z)^{1.5}$, and we also know for this case (Section 3.3.1) that $t_o = 2H_o^{-1}/3$,

$$\tau = t_o\left(1 - \frac{t_e}{t_o}\right) = 2H_o^{-1}[1 - (1 + z)^{-1.5}]/3. \tag{3.18}$$

Using this equation, the look-back time can be found. Such an equation can be derived for any value of q_o for which a relation for $R(t)$ is found. An especially simple form exists for the empty universe, $q_o = 0$, which is easy because $R \propto t$, so $R(t_o)/R(t_e) = 1 + z = t_o/t_e$, and $t = H^{-1}$. Combining these gives

$$\tau = t_o - t_e = t_o\left(1 - \frac{t_e}{t_o}\right) = H_o^{-1}[1 - (1 + z)^{-1}]$$

or

$$\tau = H_o^{-1}z/(1 + z). \tag{3.19}$$

For the case of $q_o = 0.1$, the expression is not simple. It can be derived as described by Sandage (1961) via Schmidt & Green (1983), and is: $\tau = 0.846\,H_o^{-1}\{1 - 0.0825[a(z) - a(z)^{-1} - 2\ln a(z)]\}$, with

$$a(z) = [(9 + z)/(1 + z)] + \{[(9 + z)^2/(1 + z)^2] - 1\}^{0.5}. \quad (3.20)$$

Expressions for look-back time are necessary when considering quasar evolution in Chapter 6. For most uses, it is adequate to approximate the result for $q_o = 0.1$ with the much simpler relation for $q_o = 0$.

3.7   Volumes of space

For applications in Chapters 5 and 6, calculating luminosity functions and quasar evolution parameters, it will be necessary to have expressions for the volume increment of the universe as a function of redshift. In nearby euclidean space, this is obviously $dv/dz = 4\pi d^2 dr$, with d either a proper distance or luminosity distance (the same for sufficiently small z) and dr the element of distance. With conventional units and small z,

$$dv = 8.2 \times 10^{12}\,H_o^{-3}z^2 dz\,\text{Mpc}^3\,\text{deg}^{-2}.$$

As was the case for distances and lengths, a 'proper volume' is used at larger z to take out effects resulting from expansion of the universe. This is so that luminosity functions, for example, are all normalized to physical conditions of the universe at the epoch of the relevant redshift. Use of proper volume yields equations that are equally applicable to a non-expanding universe. In general, even in a curved space, the proper volume element dv is $dv = 4\pi d^2 dr$, but the terms are now quite different conceptually. Here, d is the proper distance and dr is the proper line element along the proper distance. Both d and dr depend on the curvature of space, so must incorporate q_o. An expression has already been given for d within alternative Friedmann cosmologies (equation 3.7). Weinberg (1972), for example, shows that $dr = (c/H_o)(1 + z)^{-1}(1 + 2q_oz)^{-0.5}dz$. For alternative values of q_o, therefore:

$$dv = 8.2 \times 10^{12}\,H_o^{-3}(1 + z)^{-3}(1 + 0.2z)^{-0.5}$$
$$\times \{10z - 90[(1 + 0.2z)^{0.5} - 1]\}^2 dz\,\text{Mpc}^3\,\text{deg}^{-2},$$
$$\text{for } q_o = 0.1; \quad (3.21a)$$

$$dv = 3.3 \times 10^{13}\,H_o^{-3}(1 + z)^{-3.5}[(1 + z)$$
$$- (1 + z)^{0.5}]^2 dz\,\text{Mpc}^3\,\text{deg}^{-2}, q_o = 0.5; \quad (3.21b)$$

and

$$dv = 8.2 \times 10^{12}\,H_o^{-3}(1 + z)^{-3}z^2(1 + 0.5z)^2 dz\,\text{Mpc}^3\,\text{deg}^{-2},$$
$$q_o = 0. \quad (3.21c)$$

The quantitative difference between dv for $q_o = 0.1$ compared with that for

$q_o = 0.5$ is significant at the redshifts of quasars, being a ratio of 1.4 at $z = 0.5$, 1.9 at $z = 1$, and 2.8 at $z = 2$. Within the context of the many other uncertainties present when the dv are actually used (Chapters 5 and 6), there are not any realistic tests that can yet use observations of quasar densities to determine q_o. Translating from low redshift to high redshift quasars requires the use of evolution parameters, which can only be determined with assumed q_o.

3.8 Speculations

While the cosmologically dependent equations described in this chapter by no means exhaust all possibilities, they encompass the alternatives now considered most reasonable by observers working with quasar data. These observers are aware that other cosmologies have been proposed, some of which are drastically different from the Friedmann cosmologies used. Unfortunately, working equations analogous to those presented above are rarely given in concise form for alternative cosmologies, which somewhat discourages their use. Most observers follow the same precepts that have guided cosmological thought for centuries: use the simplest cosmology you can find that does not unquestionably disagree with the data. The equations presented in this chapter are in that spirit and will remain useful unless there is an irreconcilable confrontation over some aspects of the data. The time scale problem mentioned earlier is the primary unresolved issue. Even with the open universe ($q_o = 0.1$) that is used, a value of 100 km s^{-1} Mpc^{-1} for H_o cannot be made consistent with the ages claimed for globular clusters. For this q_o, the actual age of the universe is $0.85\,H_o^{-1}$, or 9.4×10^9 y. As stressed previously, the universe is even younger if it is flat ($q_o = 0.5$). If anything, the claims for H_o near 100 based on local measurements seem to be strengthening, and the globular cluster ages derived seem to be getting older. The primary thing still saving Friedmann cosmologies is the body of data that favors H_o near 50 on the large scale. It is not obvious where to turn if this cosmological edifice built over the past half-century should crumble. Do we introduce a cosmological constant? Do we discard the cosmological principle by assuming our local bubble of the universe is behaving atypically? Do we consider a time-varying gravitational constant? Or do we consider even more radical cosmologies? Because quasars are seen at higher redshifts than any other object, observations of quasars will play a crucial role in future efforts to reach an improved description for the true nature of the universe.

4

QUASARS AS COSMOLOGICAL PROBES

4.1   Galactic envelopes of quasars

The discovery of quasars heralded the present era of astrophysics, characterized by wide ranging investigations of every part of the spectrum, whether easily accessible or not. Observers were stimulated to open new spectral windows, primarily in the hope of finding something as extraordinary and unexpected as the quasars. None succeeded. Even when observations were pushed to X-ray wavelengths, quasars stood out. When discovered, quasars came as a stunning surprise to the small community of theoreticians who dabbled in extragalactic astrophysics. The quasars seemed so unlike galaxies that it was not clear whether their redshifts should be interpreted with the same cosmological relations that applied to galaxies. Doing so gave unbelievable answers; the quasars were just too luminous to explain. Furthermore, surprise piled upon surprise, these luminosities arose in volumes so small that luminosity variations could be seen in times of less than a year. It was fair, even necessary, to question any assumptions made for quasars, including assumptions about cosmological redshifts. I have argued at some length (Weedman 1976), so will not repeat much of it here, that this bewilderment arose as a consequence of the sequence of discovery for quasars. Had quasars been discovered initially as events in the nuclei of galaxies, the nature of their redshifts would have never been questioned. As it happened, it was only realized after the fact that identical phenomena can be observed in galactic nuclei. Once this was demonstrated, there was no further point to theoretical objections concerning redshift-derived luminosities for quasars.

The longest standing argument favoring cosmological redshifts for quasars has been this 'continuity' argument, demonstrating the unchallengeable similarities in many spectral regions between activity known in the nuclei of galaxies and the activity seen in objects discovered as 'quasars' (Yee 1980,

Kriss & Canizares 1982). The distinction between an 'active galactic nucleus' and a 'quasar' is primarily semantic, which changes depending on how carefully quasar images are examined. With diligent efforts at the telescope observers have proven that quasars are embedded within galaxies, often accompanied by other galaxies at the same redshift (Hutchings *et al.* 1984, Gehren *et al.* 1984). In several cases, absorption lines from stars in these galaxies have been seen (Boroson, Oke & Green 1982, Heckman *et al.* 1984). In most cases, the extent and luminosity of the underlying galaxy, after subtracting the quasar light, shows the characteristics of a spiral galaxy, rather than the large ellipticals which characterize radio galaxies. Correctly interpreting observations of the galaxies associated with quasars is important both to help understand the quasar phenomenon, and to isolate a distant sample of galaxies for study in their own right. Regardless of the use to which observations are put, cosmologically dependent relations enter. These relations are now summarized, as they apply to the galactic envelopes which accompany the active nuclei recognized as quasars.

4.1.1 Isophotal angular diameters

Recall from equation 3.6 that observed bolometric flux f_o from a galaxy producing luminosity L_e depends on $f_o \propto L_e d^{-2}(1 + z)^{-2}$, whereas the angular metric diameter of a galaxy goes as $(1 + z)/d$, with d the proper distance. Surface brightness, which will be denoted by I below, is measured as flux per unit area observed on the sky, usually in erg cm^{-2} s^{-1} arcsec^{-2}. So, $I \propto f\theta^{-2}$ or $I \propto (1 + z)^{-4}$. This is a severe complication for measuring galaxy diameters! The galaxy rapidly fades away with increasing redshift. Furthermore, the rate at which it appears to shrink in size depends on the surface brightness distribution across the galaxy. All parts of this distribution will fade with $(1 + z)^4$, but the apparent diameter (the 'isophotal' diameter) will depend upon the location at which the surface brightness fades below that of the measurable threshold. This threshold surface brightness will be denoted as S. It depends upon the background noise of the observing technique used. As a rule, it is not possible to detect a surface brightness S that is below 1% of the background.

As the extreme and simplest example, consider a galaxy that is a homogeneous disk with constant surface brightness within diameter D and zero surface brightness outside D. Such a galaxy would have both metric and isophotal diameters of D, as long as the surface brightness was above threshold. With increasing z, metric and isophotal diameters remain the same until suddenly the entire galaxy disappears at that z where the surface brightness falls below the threshold. In general, this is an unrealistic example, as real galaxies have surface brightness distributions that are brightest at the center and fade toward the edges. In such cases, the relation between metric and isophotal diameters as a function of z depends on the form of the galaxy surface brightness distribution. The only circumstance where this approximation of homogeneous surface brightness might be realistic would be the disk of a spiral galaxy imaged in the light of H$\text{\textsc{ii}}$ regions, which are sometimes scattered uniformly over the disk. This could be seen, for example, in interference filter imagery in the light of an emission line. For such an observation, use of relations for bolometric flux is also correct, because the total flux from an emission line is a bolometric quantity. For this example, the isophotal diameter and the metric diameter remain the same at all z, until the entire galaxy suddenly disappears from detection when I drops below S.

An important issue concerning quasars is that of the form of the host galaxy, so it is necessary to compare host galaxies at different z, in which case we are forced to deal with the complex relations for galaxy isophotal diameters. As will be seen, these depend on the form of the intrinsic $I(\theta)$ within the galaxy. Existing empirical determinations of galaxy surface brightness distributions have been made by optical magnitude measurements of the light in the continuous spectrum, so they arise from $f_{\lambda o}$. To discuss these models, monochromatic surface brightness in the observer's frame, I_λ, is used rather than bolometric I. Because $f_{\lambda o} = L_{\lambda e}/4\pi d^2 (1 + z)^3$, $I_\lambda \propto f_{\lambda o} \theta^{-2} \propto (1 + z)^{-5}$, rather than $I \propto (1 + z)^{-4}$ for bolometric I. In most literature discussing surface brightnesses of galaxies (Petrosian 1976, Tinsley 1976), such brightnesses measured from magnitudes are stated to scale with $(1 + z)^{-4}$, rather than the $(1 + z)^{-5}$ given here. This is not an inconsistency. Recall that there are five powers of $(1 + z)$ involved. One is for change in photon energy, one for change in unit time, two for change in unit area of the galaxy, and one for change in bandpass. This last is the source of the difference in describing how surface brightness changes with redshift. Historically, the bandpass term has been incorporated into the K-correction, which includes both the bandpass term and the correction for observing at different parts of the emitted spectrum at different redshifts. I am not using tabulated K-corrections, but instead use relations referred to the intrinsic spectrum of the source (Section 3.5.2). As a consequence, an extra power of $(1 + z)$ enters when describing observations of this intrinsic spectrum. This approach is particularly useful when dealing with the surface brightness problem, because the term which scales the surface brightness profile of a galaxy is the central surface brightness that would be seen at negligible redshift, denoted I_λ (o,0). This gives the intrinsic central surface brightness that would be seen by an observer close enough for no cosmological effects to enter. Observationally, such intrinsic I_λ (o,0) can be determined at any wavelength by obtaining the spectra of local galaxies, observed in the ultraviolet.

To make illustrative isophotal diameter calculations, real data which are now available for quasar envelopes will be used. To accommodate galaxies with different stellar contents, which would thereby have different spectral shapes, representative spectra for galactic envelopes are given in Table 4.1. The entries display the ratio of surface brightness at any wavelength to that at 6500 Å, which is the observing wavelength for which calculations are to be made. The continuum spectral forms for different galaxy types, ignoring

Table 4.1. *Change of galaxy surface brightness with wavelength*

| Emitted wavelength λ_e (Å) | Surface brightness at λ_e/surface brightness at 6500 Å | | |
	Old-star galaxy	Spiral galaxy	Young-star galaxy
6500	1	1	1
6000	1.03	1.03	1.09
5500	0.97	1.11	1.19
5000	0.89	1.13	1.34
4500	0.82	1.19	1.58
4400	0.72	1.11	1.53
4300	0.60	1.05	1.50
4100	0.57	1.03	1.67
4000	0.49	1.01	1.67
3900	0.28	0.91	1.62
3700	0.33	0.79	1.42
3500	0.27	0.65	1.37
3300	0.25	0.53	1.38
3200	0.24	0.48	1.41
3100	0.16	0.45	1.44
2900	0.12	0.41	1.55
2800	0.083	0.41	1.63
2700	0.060	0.41	1.72
2600	0.036	0.41	1.82
2500	0.029	0.41	1.95
2200	0.028	0.41	2.41
2000	0.028	0.44	2.81
1700	0.024	0.42	3.35
1500	0.019	0.37	3.65
1200	0.017	0.33	4.11

emission lines, are taken from Coleman, Wu & Weedman (1980). The old-star galaxy is the mean of nuclear bulges for M31 and M81; spiral galaxy is an Sbc; young-star galaxy is the blue irregular galaxy NGC 4449, now containing extensive star formation. As observations are made at progressively higher redshift, the intrinsic wavelength observed changes as $\lambda_e = 6500(1 + z)^{-1}$.

Most elliptical galaxies have their surface brightness distributions modelled by a relation (de Vaucouleurs 1959) that describes the surface brightness as a function of distance θ from the center by $I_\lambda(\theta) = I_\lambda(o,z) \exp - (\theta/r)^{0.25}$. $I_\lambda(o,z)$ is the central surface brightness and r is a parameter determining the scale size of the galaxy; galaxies that are larger in the sky have larger r. This r is defined in units of an angular radius, but it describes that radius in the galaxy at which the surface brightness drops to a specified fraction of the central surface brightness. Consequently, r is a metric quantity, and it changes with z according to equations 3.17a and b. Some physical length in the galaxy gives rise to the r and could be found from equations 3.17a and b. $I_\lambda(o,z)$, however, is a surface brightness and so decreases with $(1 + z)^5$; $I_\lambda(o,z) = I_\lambda(o,0)(1 + z)^{-5}$. The constant $I_\lambda(o,0)$ can be evaluated at any z sufficiently low that the cosmological effects on surface brightness are negligible. The notation does not literally require determining the central surface brightness of galaxies at zero redshift. The λ_e for which $I_\lambda(o,0)$ is taken from Table 4.1 depends on the redshift. For observations made at any λ_o, $\lambda_o = (1 + z)\lambda_e$. This incorporates the correction for the intrinsic spectral shape.

The definition of isophotal diameter $2\theta_i$ is such that the observer's task is to locate that value of θ_i at which the $I_\lambda(\theta_i)$ has the limiting detectable value, here denoted S. S is determined by the conditions of the measurement, not by the properties or redshift of the galaxy being scrutinized. The objective is to find the $\theta_i(z)$ at which $S = I_\lambda(o,z) \exp - (\theta_i/r)^{0.25}$. Using the $I_\lambda(o,z)$ as already defined and the metric relations for the physical scale factor G corresponding to r, relations for $\theta_i(z)$ can be found for different q_o. G defines the proper radius of the galaxy in kpc at which the surface brightness is e^{-1} of the central surface brightness. Then,

$$\theta_i'' = (1 + z)^2 \{\ln [(1 + z)^{-5} I_\lambda(o,0) S^{-1}]\}^4 (6.88 \times 10^{-4} H_o G)$$
$$\times \{10z - 90[(1 + 0.2z)^{0.5} - 1]\}^{-1}, \quad q_o = 0.1; \quad (4.1a)$$

$$\theta_i'' = (1 + z)^2 \{\ln [(1 + z)^{-5} I_\lambda(o,0) S^{-1}]\}^4 (3.44 \times 10^{-4} H_o G)$$
$$\times [(1 + z) - (1 + z)^{0.5}]^{-1}, \quad q_o = 0.5. \quad (4.1b)$$

The alternative galactic form which is most important is that for a spiral galaxy observed in the light of the continuum. Such a surface brightness distribution is represented as an 'exponential disk', which is described by $I_\lambda(\theta) = I_\lambda(o,z) \exp - \theta/r$. Proceeding as before, this leads to:

$$\theta_i'' = (1 + z)^2 \ln [(1 + z)^{-5} I_\lambda(o,0) S^{-1}](6.88 \times 10^{-4} H_o G)$$
$$\times \{10z - 90[(1 + 0.2z)^{0.5} - 1]\}^{-1}, \quad q_o = 0.1; \quad (4.2a)$$

and

$$\theta_i'' = (1 + z)^2 \ln [(1 + z)^{-5} I_\lambda(o,0) S^{-1}](3.44 \times 10^{-4} H_o G)$$
$$\times [(1 + z) - (1 + z)^{0.5}]^{-1}, \quad q_o = 0.5. \quad (4.2b)$$

Real spiral galaxies may be combinations of the two profiles (Kent 1984), with an $r^{0.25}$ component dominating the central bulge and the exponential profile controlling the outer portions. Because both $I_\lambda(o,0)$ and the scale parameter G are free parameters for a profile fit, there is no large error in forcing a given galaxy to fit one or the other profile shapes.

The galactic envelopes of quasars have been observed to redshifts of about 0.5. These envelopes are sufficiently small that it is difficult to choose whether they fit better with an $r^{0.25}$ or an exponential disk model. These models differ most significantly within the inner portions of a galaxy, with the 'elliptical' law peaking much more strongly to the galactic center. Of course, if a quasar resides in the center, then the peak of the galaxy's starlight distribution cannot be observed, and outer parts of the galaxy are used to fit a profile. Using the fits of Malkan, Margon & Chanan (1984) to quasar envelopes, the parameters for profiles would be, for an elliptical model, $I_{4400}(o,0) = 9.3 \times 10^{-15} \, \mathrm{erg \, cm^{-2} \, s^{-1} \, Å^{-1} \, arcsec^{-2}}$ and $G = 2 \times 10^{-3} \, \mathrm{kpc}$. For an exponential disk, $I_{4400}(o,0) = 2.9 \times 10^{-17} \, \mathrm{erg \, cm^{-2} \, s^{-1} \, Å^{-1} \, arcsec^{-2}}$ and $G = 6.4 \, \mathrm{kpc}$. These are values to which the $I_\lambda(o,0)$ in Table 4.1 are scaled. These parameters for quasar envelopes are similar to those for normal galaxies. Considering entire galaxies as tertiary distance indicators, this result is evidence that, at least to $z \approx 0.5$, quasars are at their cosmologically derived distances. That is, quasars obey the redshift–distance relation just as well as normal galaxies. Supplementary proof is in the frequent association of quasars with other galaxies at the same redshift.

The scale length for an e^{-1} change in surface brightness is so small for the elliptical profile that it is not a realistically observable parameter. What is done in practice is to observe the I_λ at some θ, and fit the profile to give the formal parameters $I_\lambda(o,0)$ and G. Often, the θ which is quoted is the 'effective' angular radius, or that which includes half of the total light from the galaxy. In terms of proper lengths, this effective radius is $3.5 \times 10^3 \, G$. For an exponential disk, by contrast, the central surface brightness is an observable quantity for nearby galaxies because it does not fall off too fast in the inner regions. As a result, the quoted scale size is normally G.

Once we have an idea what the envelopes of nearby quasars look like, the cosmological equations for isophotal diameter can be used to predict the appearance at much higher redshifts. A variety of other parameters enter these calculations. An ultraviolet spectrum must be assumed so the $I_\lambda(o,0)$ are known when the observer's wavelength is greatly different from the rest wavelength in the quasar. An old stellar population has very little ultraviolet flux compared with that in the visible, whereas young, hot stars are most conspicuous in the ultraviolet. The intrinsic spectra that will be used for sample calculations are those in Table 4.1, taken from observed spectra of nearby galaxies. An old stellar population, a 'spiral' population, and a young population could all be considered.

Another necessary decision is the limiting surface brightness S_λ, which is detectable for defining the limiting galactic isophote. For ground based observations, this is set primarily by the background sky limit. From space, it will probably be set primarily by the detector noise limit. In either case, the limiting detectable surface brightness is fainter than the background brightness which is seen, because the galaxy is extra light superimposed on that background. How much fainter the galaxy can be and still be detected depends on the uniformity of the background. It is the fluctuations rather than the absolute amount of the background that confuse the detection of a real object on this background. In the most optimistic case, it should prove possible to detect a surface brightness of 1% of the background. I have chosen an observing wavelength of 6500 Å for all calculations, because that is the wavelength of maximum efficiency for CCD detectors. At this wavelength, an optimistic case, corresponding approximately to the 1% detectability from either ground or space, gives a limiting detectable surface brightness of 10^{-19} erg cm^{-2} s^{-1} Å^{-1} arcsec^{-2}. To illustrate a limit realistically achieved, calculations will also be given with this limit increased a factor of 10 to 10^{-18}.

Two alternative Friedmann cosmologies are already included in equations 4.1 and 4.2. Now is an opportunity to note that measures of galactic isophotal diameters give a chance of determining if redshifts are really caused by the expansion of the universe (Chapter 3). Within the relations for surface brightness are included three powers of $1 + z$ that enter because of the assumption that the universe is expanding. Two powers enter because the unit area of the galaxy observed goes as a proper length squared, and proper lengths relate to observed lengths with a term of $1 + z$. Also, a single power of $1 + z$ enters from the flux–luminosity equations because the duration of a unit time changes as $1 + z$. The remaining two powers of $1 + z$ in the surface brightness portion of equations 4.1 and 4.2 do not require the expansion of the universe as they arise from the change in unit wavelength and the change in photon energy by $1 + z$, both of which would arise from any redshift effect. Using the same space curvature (existence of which does not require an expanding universe) for a universe with $q_0 = 0.1$ or 0.5, equations 4.1 and 4.2 are modified to determine the isophotal diameters for a non-expanding universe first by changing the $(1 + z)^{-5}$ terms to $(1 + z)^{-2}$. Secondly, the first term in each equation, $(1 + z)^2$, arises from the metric angular size. That also contains a power of $1 + z$ that arises from expansion (Section 3.5.3), so this first term must be altered to $1 + z$.

In Table 4.2 are summarized the results of calculations with a variety of combinations of the parameters discussed, starting with the empirically determined $I_\lambda(o,0)$ and G for the galactic envelopes of low redshift quasars. If we are interested in the redshift limit to which galactic envelopes of quasars can be observed, the relevant results are those for diameters of 2", approximately the ground based resolution limit, or 0.2", approximately the resolution limit for imagery from space. Because the galactic envelopes of

Table 4.2*a*. *Isophotal diameters for spiral galaxy quasar envelopes*[a]

Redshift	Diameter 2θ; for $S = 10^{-18}$ erg cm^{-2}s^{-1}Å^{-1}arcsec^{-2}	Diameter 2θ; for $S = 10^{-19}$ erg cm^{-2}s^{-1}Å^{-1}arcsec^{-2}
0.05	46"	76"
0.1	22	39
0.2	11	21
0.3	6.8	14
0.4	4.8	11
0.5	3.1	8.8
0.6	2.1	7.2
0.7	1.0	5.8
0.8	—	4.6
0.9	—	3.5
1.0	—	2.6
1.1	—	1.9
1.2	—	1.4
1.3	—	0.9
1.4	—	0.5
1.5	—	0.16
1.6	—	—
1.7	—	—

[a] Galaxy form same as seen around low redshift quasars, with exponential disk of scale length $G = 6.4$ kpc and central surface brightness at 6500 Å of 2.6×10^{-17} erg cm^{-2}s^{-1}Å^{-1}arcsec^{-2}.

quasars are similar to other galaxies (Yee 1983), these results are also representative for galaxies not containing quasars.

It is clear from Table 4.2 that quasar envelopes should not be seen with any technique much beyond $z = 1$ if those envelopes are like nearby quasars, with the 'normal' scaling parameters and the spiral ultraviolet spectrum. It

Table 4.2b. *Isophotal diameters for envelope galaxies of various types*[a]
($q_o = 0.1$, $S = 10^{-19}$ erg cm^{-2} s^{-1} Å^{-1} arcsec^{-2})

Redshift	$2\theta_i$ for old-star galaxy	$2\theta_i$ for young-star galaxy	$2\theta_i$ for most optimistic galaxy
0.05	82″	72″	123″
0.1	42	37	65
0.2	22	20	36
0.3	15	14	26
0.4	11	11	21
0.5	8.6	8.9	18
0.6	6.8	7.7	16
0.7	4.5	6.3	14
0.8	3.7	5.3	13
0.9	2.8	4.6	12
1.0	2.1	4.0	11
1.1	0.8	3.5	10
1.2	0.03	3.1	9.6
1.3	—	2.8	9.1
1.4	—	2.5	8.8
1.5	—	2.2	8.4
1.6	—	2.0	8.2
1.7	—	1.7	7.9
1.8	—	1.6	7.7
1.9	—	1.3	7.4
2.0	—	1.1	7.2
2.1	—	0.9	7.0
2.2	—	0.7	6.8
2.3	—	0.6	6.7
2.4	—	0.4	6.5
2.5	—	0.1	6.3
2.6	—	—	6.2
2.7	—	—	6.0
2.8	—	—	5.8
2.9	—	—	5.6
3.0	—	—	5.4

[a] Old-star and young-star galaxies have same parameters as galaxies observed around quasars, as in Table 4.2a, except for changing stellar populations. Most optimistic galaxy takes central surface brightness at 6500 Å of 7×10^{-16} erg cm^{-2} s^{-1} Å^{-1} arcsec^{-2} with $G = 6.4$ kpc and young star population.

makes little difference whether $q_o = 0.1$ or 0.5. Only if the universe is not expanding do normal parameters yield detectable galaxies at $z \sim 2$, and even then only for the optimistic detector limit. The results in Table 4.2 are discouraging for hopes to study distant galaxies. On the other hand, we know nothing about the nature of galaxies at high redshift, and stars had to form

Table 4.2c. *Isophotal diameters for spiral envelopes with different cosmologies*[a]

$(S = 10^{-19} \, \mathrm{erg \, cm^{-2} \, s^{-1} \, \mathring{A}^{-1} \, arcsec^{-2}})$

Redshift	$q_o = 0.1$, but universe not expanding	$q_o = 0.5$, universe expanding
0.05	74"	77"
0.1	37	40
0.2	19	22
0.3	13	15
0.4	10	12
0.5	8.0	9.7
0.6	6.3	8.2
0.7	5.4	6.6
0.8	4.5	5.4
0.9	3.8	4.2
1.0	3.3	3.2
1.1	2.9	2.4
1.2	2.5	1.7
1.3	2.3	1.1
1.4	2.1	0.7
1.5	2.0	0.2
1.6	1.8	—
1.7	1.7	—
1.8	1.6	—
1.9	1.5	—
2.0	1.4	—
2.1	1.3	—
2.2	1.3	—
2.3	1.2	—
2.4	1.1	—
2.5	1.1	—
2.6	1.0	—
2.7	0.9	—
2.8	0.9	—
2.9	0.8	—
3.0	0.8	—

[a] Same galaxy as used for Table 4.2a is considered here with alternative cosmological assumptions.

prolifically at some epoch. So it might be hoped that at some high redshift quasars will be observed surrounded by galactic disks made very luminous by star formation. I have determined the 'most optimistic' case by taking the brightest known local disk of star formation, which has $I_{6500}(o,0) \approx 7 \times 10^{-16}$ (Weedman & Huenemoerder 1985), scaling it up in the ultraviolet with the young star spectrum from Table 4.1, and allowing it to be as large in G as the known quasar envelopes. For this case, it might be possible to detect the envelope to $z > 3$ from the ground. That such envelopes have not been seen indicates that this case is far too optimistic. Whether large galactic envelopes are ever found at high redshift will be a very instructive result regarding the epoch of formation for the galaxies associated with quasars.

4.2   Absorption lines

Once it is established that quasars are at cosmological distances, they are useful as probes of the intervening universe-like background searchlights – even if the quasar luminosity sources are not understood. In this context, comments concerning absorption lines in quasar spectra would be out of place if those lines are associated with material in the quasar itself. Whether absorption lines are arising intrinsically to the quasar or in intervening material was the initial question which aroused interest in absorption features. The intensity of the controversy was fuelled by its implications for the nature of the quasar redshifts; if the lines are associated with quasar material, they provide no evidence for cosmological quasar redshifts. Absorption materials seen at redshifts close to that of the quasar which arise from intervening galaxies would require cosmological redshifts for the quasar. While all parties are not satisfied with the compromise which has resulted, it now seems that both intrinsic and intervening interpretations are correct, depending on the lines concerned.

There is one category of quasars, including a few per cent of all quasars, for which there is consensus that the dominant absorption features are intrinsic to the quasar. These are the broad absorption line quasars, or BAL quasars (Turnshek 1984). These show strong resonance absorption lines, primarily Ly α and C IVλ1550, which are very broad, extending continuously from the emission line redshift blueward to velocities exceeding 10^4 km s^{-1}. Such lines are often described as P Cygni profiles, by analogy to similar features seen in stars that are ejecting material toward the observer. For various reasons, it is not surprising to see such material flowing out of a quasar at such velocities, so interpretations of these BAL quasars are built around scenarios describing gas flows near the quasar. More is said about BAL quasars in Chapter 9 on quasar structure.

More controversial are the discrete, narrow absorption line features found with great frequency in quasar spectra (Weymann *et al*. 1979, Sargent *et al*. 1980). Sometimes, a pattern of lines can be identified with different features (usually Ly α and C IV) at the same redshift. More common are the large numbers of single lines invariably found blueward of Ly α emission, termed the 'Ly α forest'. Many quasar spectra are completely broken up shortward of Ly α by these features. It takes very high spectral resolution just to count the lines, because they typically have widths of order 10 km s^{-1}; worse resolution causes different features to blend. This is why one finds a new round of quasar absorption line studies whenever instrumental improvements are made at telescopes.

In part, it is the narrow widths of these lines that argue against their intrinsic association with the quasar. It is difficult to understand how cloudlets with internal velocity dispersions of 10 km s^{-1} could exist close to the environment of a quasar where gas velocities are characteristically thousands of km s^{-1} and where all velocities of the cloudlets should not necessarily be toward the observer. There are other arguments against the hypothesis that these narrow lines arise in gas being ejected toward us from a quasar. These include the difficulty of explaining how ejection could lead to a number density of clouds smoothly distributed with velocity and showing line strengths having no correlation with velocity. While one could produce *ad hoc* quasar models with cloud structures that lead to the lines seen, it is expected that such lines should be seen from intervening material between the quasar and the observer. Consequently, it is natural and reasonable to interpret the lines which are seen as arising from such material. Then, the characteristics of these narrow quasar absorption lines provide valuable information on the gaseous component of the universe between observer and quasar.

That absorption lines in quasar spectra can be produced by very small amounts of material is shown by quantitative considerations. The formation of absorption lines is a problem thoroughly treated by radiative transfer theory. The simplest case arises for optically thin lines, for which the only line broadening is caused by the Doppler width of the line. This width is a measure of random atomic motions in the absorbing cloud, only a few km s^{-1} in the Ly α forest lines. Using classical theory describing absorption via f values, the equivalent width in angstroms of a weak Ly α line is $W = 0.5 \times 10^{-14} N_0 \mathrm{d}s$. Here, N_0 is the number cm^{-3} of neutral hydrogen atoms, almost all of which are in the ground state. This is only some fraction of the total hydrogen density N_H ($N_H = N_0 + N_p$). This fraction may be very small if the cloud is ionized, as N_H is then primarily N_p. The term ds is the path length through the cloud in cm. Because of the resulting units, the product $N_0 ds$ is the 'column density', or the total number of atoms cm^{-2} in the observer's line of sight through the cloud. With high resolution spectroscopy, absorption features with equivalent widths less than 1 Å can be detected. This means observations are sensitive to the presence of clouds having column densities N_0 in neutral hydrogen of $\sim 10^{14}$ cm^{-2}. This discussion refers to the equivalent width intrinsic to the quasar rest frame. Recall from Chapter 3 that the observer sees an equivalent width that is greater by $1 + z$.

Typical interstellar matter densities N_H in galactic disks are 1 cm^{-3} rising to 10^4 cm^{-3} in bright nebulae but below 10^{-6} cm^{-3} in intergalactic space. Consider a low density shred or cloud of material, taking $N_0 = 10^{-3}$ cm^{-3}, which might be an isolated cloud or might be in the outer reaches of a galaxy. Were this cloud spherical, it would need a radius of only 1.5×10^{-2} pc to have a column density of 10^{14} cm^{-2}, yielding a total mass in the cloud of 3×10^{-10} solar masses.

If small, isolated clouds of material of this size fill the universe, there is no way to see them other than with absorption lines. They cannot support star formation and are of such low density that recombination emission lines are insignificant. Even though measurable absorption lines can be produced by very small clouds, there may be much thicker column densities. Were the disk of a galaxy seen in absorption, with $N_0 \approx 10^{21}$ cm^{-2}, the profile would become optically thick. Line depth would reach zero intensity, and the damping wings of a Voight profile would appear, broadening the line compared with its intrinsic Doppler width. By searching for such profiles, the disks of unobserved galaxies can be probed (Wolfe *et al.* 1986).

Looking for absorption lines in quasars is a necessary probe, therefore, for such constituents of the universe. Ly α absorption only occurs from neutral hydrogen, so the relevant column density is that for N_0. If the clouds are primarily ionized, then the actual column densities could be much higher than that for N_0, raising the masses correspondingly. In fact, even if there are no ionizing sources within the clouds, the general radiation field from background sources – primarily quasars – should keep the clouds quite highly ionized. The ratio N_0/N_H is probably less than 10^{-3} (Black 1981). For an ionized cloud, therefore, the column density of N_0 may not be much greater, perhaps even less, than the column density of a highly ionized ion whose neutral parent has a much smaller abundance than hydrogen. This is the reason that absorption features from C IV can compare in strength with Ly α features even though carbon is much less abundant than hydrogen.

Learning whether narrow absorption line systems are strictly Ly α or also contain features from heavier elements is an important objective of any absorption line study. Making this determination requires careful correlations among the many absorption features seen. If heavy elements are present, it is a demonstration that the gas has been processed and so has at one time been part of a galaxy in which star formation and evolution have occurred. This simplistic viewpoint leads to a first-order distinction between categories of narrow absorption lines. The Ly α forest lines, not accompanied by analogous absorption in other features, are attributed to a population of intergalactic clouds of primordial material, never part of any galaxy (Sargent *et al.* 1980). This is a necessary interpretation because the lines are so abundant that galactic haloes would have to be defined as filling the universe if the forest lines are attributed to haloes. On the other hand, those lines that are seen both in hydrogen and carbon are attributed to absorption by galactic haloes (Weymann *et al.* 1979). That galactic haloes indeed produce absorption has been demonstrated using the halo of our own Galaxy (Savage & Jeske 1981), and the column density of N_H through a galaxy is sufficiently large (typically 10^{21} cm^{-2} for a spiral disk) that absorption far out in the disk or the halo is to be expected. Major questions are how far out the haloes extend, how they depend on the associated galaxy type, what are the relative ionic abundances, and what are the values and sources of the absorption line widths. All of these questions about haloes cannot be answered until sample haloes of nearby galaxies can be probed with the ultraviolet absorption lines, primarily Ly α and C$_{IV}$, silhouetted against background quasars.

There remains much unexploited potential in the examination of the Ly α forest lines. If these are really showing residual remnants of primordial gas, this is our only opportunity to probe a non-galactic constituent of the universe. It is exciting to determine how such leftover shreds are distributed through the universe, both with epoch and location. Consider the cosmological relations involved. Assume first that such clouds are distributed uniformly with co-moving density ρ at all times in the universe, and take a representative cloud radius to be G. In the local universe, the number of clouds dN seen along a path interval ds through space would be $dN = \rho \pi G^2 ds$. For local euclidean space, $ds = c H_o^{-1} dz$, so $dN = \rho \pi G^2 c H_o^{-1} dz$. To get an analogous expression at cosmological distances, the path interval ds has to be a proper length and G has to be a proper length; ρ is already a proper density. In the discussion of co-moving volume elements in Chapter 3, it was explained that the proper length interval $dr = c H_o^{-1} (1 + z)^{-1} \times (1 + 2q_o z)^{-0.5} dz$. In the discussion of angular diameters, it was pointed out that a proper length G would appear to the observer to have length $(1 + z)G$ in an expanding universe. Inserting these modifications,

$$dN(z) = \rho \pi G^2 c H_0^{-1}(1 + z)(1 + 2q_0 z)^{-0.5}dz. \qquad (4.3)$$

A distribution of Ly α forest lines could show the distribution of clouds along a path through most of the universe if the lines can be followed in a single quasar from the local epoch ($z \approx 0$) all the way to the quasar redshift. The fundamental question of interest is whether the cloud density ρ or size G is a function of redshift. Has this population of clouds evolved? Ground based studies can already consider the distribution at early epochs, $z \gtrsim 2$. From studies done so far, the normalizing product $\rho \pi G^2 c H_0^{-1}$ seems consistent among different quasars at high redshift, being 20 to 30 (Sargent *et al.* 1980). There are indications that absorbing material is more common at higher redshifts in a few carefully studied quasars, thereby providing a hint of another evolving component of the universe (Atwood, Baldwin & Carswell 1985). The density of clouds observed is a negligible fraction of the mass in the universe so does not affect Ω and q_0.

Close groupings of quasars or separate images of lensed quasars (Section 4.3) can be used to see how the distributions differ for slightly different paths through space (Foltz *et al.* 1984, Sargent, Young & Schneider 1982). So far, this has set upper limits to the sizes of the Ly α forest clouds of order 10 kpc. To compare $\rho(z)$ for clouds, fundamentally similar clouds can be defined using line width as the criterion, rather than absorption line depth, so some of the uncertainties of varying ionization degrees in the clouds can be avoided. The answers that will come from these various studies are unpredictable, which is why they hold such potential for interesting new observations, especially if lower redshifts are probed. The utility of the quasars is simply in providing a backlighting source to probe these quiescent clouds. The brighter the quasars, the more use they are in yielding high resolution spectra. This should give major incentive to those quasar surveyors (Chapter 2) who are capable of finding bright quasars. For mapping the Ly α forest throughout the universe, all of the bright quasars available are important.

4.3   Gravitational lenses

As was emphasized in Chapter 3 on cosmological concerns, we are fundamentally baffled by the question of the form in which most of the mass of the universe is found. Velocity dispersions of non-cosmological motions for galaxies are so high that they imply ten times as much mass present as is visible in the galaxies' stellar composition. If a flat universe ($\Omega = 1$) is accepted to please the esthetic desires of inflationary cosmologies, this shortfall is a factor of 100. The difficulty is that most of the mass in the universe seems to be dark, so the conventional techniques of astronomy –

trying to see the light – are useless. Such matter can only be revealed by gravitational effects. If there is no visible matter in the vicinity of the dark matter, how could the gravitational effects be observed? The only way is by the effect of gravity on light passing by, an effect detectable as a consequence of gravitational lensing.

Studying the consequences of gravitational lensing is a complex discipline in itself, in which quasars have unique and very exciting potential for use. The quasars provide excellent backlighting probes of everything in between them and us. If, anywhere on the journey of the light from the quasar to the observer, this light passes a massive object, the path of the light is affected. Whether the effect is significant enough for detection by the observer is the question of interest.

There are several known cases of quasars whose light has been deflected by intervening masses such that observers see multiple images of a given quasar. Gravitational lensing is most easily demonstrated by finding a pair of quasars, each of which has the same redshift, but which are separated in the sky by only a few arcseconds. If the spectra are sufficiently identical, it can be argued that the two images really are of a single quasar. In a few cases, a galaxy and/or galaxy cluster has been seen that provides the necessary mass to explain gravitational lensing (Young *et al.* 1980). For the first such configuration discovered, the lens has been thoroughly modelled, and provides an excellent free-standing proof of a situation with a quasar at truly cosmological redshift far beyond an intervening galaxy (Gorenstein *et al.* 1984).

Describing the images produced by a gravitational lens is very complex, because galaxies and clusters do not image as if they were point masses. Only the most elementary approximations of lensing will be discussed here, primarily to point out some of the potential of lensing for learning more about the universe.

That gravitational lensing should occur for celestial objects had been realized for a long time, and various predictions were made based on this expectation (Einstein 1936, Refsdal 1964, Barnothy & Barnothy 1972, Press & Gunn 1973). Only with the first discovery of a quasar that is actually lensed (Walsh, Carswell & Weymann 1979) did many efforts begin to explain the observed consequences of lensing in realistic cases. Interpretations are complex, not because the lensing process is complex, but because the distribution of matter in the lens can take many forms. Being gravitationally lensed affects a quasar in two ways: (*a*) Angular size and shape of the image is changed. This effect can cause image splitting and makes it possible to detect lensed quasars by searching for two images of one quasar. (*b*) The quasar is brightened if the light from the entire split image is compared with that which would have been seen in the absence of lensing. Even if image splitting is not resolvable, the image brightening occurs.

The basics are illustrated by a schematic of lensing in the simplest possible case, that of a point source quasar lensed by a point mass which is precisely aligned with the observer in euclidean space. The geometry is as in Figure 4.1 Rays from the quasar are deflected by the mass at an angle δ given by $\delta'' = 1.2 \times 10^{11} \, Mb^{-1}$, where the deflecting mass M is in solar masses, and the impact parameter b is in cm. The observer would think the incoming rays derived from the directions shown, so would see images of the quasar in those directions. This special case has such symmetry that any ray going past the lensing mass at distance b would be similarly deflected, so the observer would see an image of the quasar which was a ring whose diameter depends on the relative placement of o, M and q. If M were precisely midway between o and q, the imaged ring diameter, equivalent to the amount of image 'splitting', would be $2\alpha = \delta$. For other placements, α never exceeds δ. In the case to be considered below of imaging by point masses in nearby galaxies, $oq \gg oM$ so $\alpha = \delta$. Not only does the lensing mass change the size of the quasar image in this way, but there is more light in the ring than would have been in an unlensed quasar image, because the deflecting mass has 'collected' rays over a larger solid angle than would otherwise reach the observer. As a result, the quasar is intensified.

Such a simple case can never occur for quasar images actually seen as multiple. Any mass large enough to cause detectable splitting is not a point mass, but is a distributed mass such as a galaxy or cluster of galaxies. Then the form of the re-imaging depends both on the precise location of M relative to o and q as well as the distribution of mass within the lensing object. Highly complex configurations and results have been seen and worked out. Nevertheless, some important concepts can still be illustrated quantitatively with the simple relations given. For example, the actual re-image of the quasar will have a few components whose maximum splitting will be about

Figure 4.1. Geometry of simple gravitational lensing by a point mass.

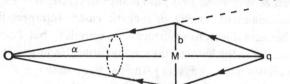

the same as the ring diameter above. The reason is that the lens cannot be far off line oq, and as the lens is displaced slightly relative to this line the ring breaks up into components whose intensities change rapidly with the displacement, but their relative splittings remain roughly as the ring diameter in the idealized case. The means that easy estimates can be made for the amount of splitting that might be observed for real lenses. For example, an elliptical galaxy containing $10^{12} \, M_\odot$ within an impact parameter of 10 kpc would give $\delta = 4''$.

This is enough to predict that lensed splitting of quasar images by real galaxies might be observable. Cosmological considerations complicate the interpretation only in the sense that distances within the geometry of Figure 4.1 have to be replaced by proper distances, and length b by a proper length, using the relations in Chapter 3. The feasible complexities of the mass distribution in the lens are much harder to accommodate, because the mass in galaxies and clusters of galaxies is not distributed in simple ways.

There are plenty of objects in the universe – stars and their remnants – that are effectively point masses as far as their lensing effects are concerned. Image splitting of a detectable amount by individual stars in galaxies is not feasible for reasons more subtle than the simple expression for δ. Consider $10 \, M_\odot$ within a canonical stellar radius of 10^{11} cm; taking this radius as the minimum possible b yields $\delta = 10''$. But stellar remnants, such as white dwarfs, neutron stars or black holes, can be much more compact than this, implying that δ could become very large even for a few $M_\odot$ as the allowable impact parameter decreases. The reason this cannot occur is that the lensing mass must be significantly larger in angular size than the quasar that is being re-imaged. If only a small part of the quasar is lensed, the net effect would be unnoticeable. While it is appropriate to consider quasars as point sources when deriving lensing effects by galaxies that are much larger in physical size than the quasar, this is not appropriate for lensing by objects of stellar size or smaller. The smallest region in a quasar is that in which the observable continuum arises, which is typically no smaller than 0.01 pc (Chapter 9). At a representative redshift of $z = 2$, the equations of Chapter 3 with $q_0 = 0.1$ yield an apparent metric angular diameter for this continuum region of $1.3'' \times 10^{-6}$. Because the metric diameter changes little with redshift above $z \sim 1$ (Table 3.1), this is a usable value for virtually any quasar that might be lensed.

Consider the angular diameter of a star within a nearby galaxy, a star which might be lensing a quasar. For any other galaxy, the distance to the star exceeds 1 Mpc, so a star of radius 10^{11} cm would subtend $<10^{-8''}$, substantially less than the quasar diameter. This means that an impact parameter for useful lensing must greatly exceed the radius of even a normal star, and must be just as large regardless of how the mass of the star is compressed. So it makes no difference for the lensing whether the mass is distributed in a star or a compact stellar remnant. The lensing to be of meaning (so the b compares to the apparent quasar size) must occur with $b \gtrsim 10^{13}$ cm, yielding $\delta < 0.1''$ for even the closest external galaxies and lensing objects of 10 $M_\odot$.

Stars in the halo of our own Galaxy would be much closer and would subtend angular sizes exceeding those of a quasar continuum region. It is not thought that such stars significantly exceed 1 $M_\odot$, so the maximum image splitting would be of order $1''$. There is other reasoning why it is very unlikely to detect quasar lensing by galactic halo objects, primarily the improbability of any lines of sight out of the galaxy intercepting a star. The same disadvantage applies to observations through the haloes of other galaxies, except that all paths through such haloes can be monitored simultaneously and for many different galaxies. Detecting lensing splitting from objects in our own halo is made more difficult by the transverse velocity of such objects; a star at distance 1 kpc moving at 300 km s^{-1} would eclipse a quasar for only an hour.

While these numbers tell us that stars or compact stellar remnants cannot be detected as lenses because of quasar image splitting, such objects could still produce a substantial change in the quasar brightness by the gravitational imaging. This intensification can be substantial. Amplification of the quasar brightness is significant only when the lensing object is nearly aligned with the quasar. In that case, the amplification factor is approximately $10^{-4} \beta^{-1}(M/d)^{0.5}$, where β'' is the angular displacement of the lens from the line of sight to the quasar, M is the lens mass in $M_\odot$ and d is the lens distance in Mpc (Canizares 1981). Considering an angular displacement comparable to the quasar angular sizes of $10^{-6\prime\prime}$ deduced above, this relation shows that significant amplification occurs for lenses of a few solar masses at distances of many Mpc. This realization means that the presence of dark, concentrated objects in the universe with a wide range of masses might be manifested by intensification of quasars (Canizares 1982).

How could we ever be confident that any given quasar is lensed if no splitting is seen? Probably, we will never be, but quasars intensified by small lenses should show certain systematic differences from non-lensed quasars. As a result, a class of such intensified quasars might be definable statistically. The differences are in variability characteristics and in equivalent widths of the emission lines. Small, dark lenses in galactic haloes should be moving and so would pass in front of the background quasar. The quasar intensification would only last times of order months to years, and the variability would be a simple turn-on, turn-off phenomenon. Realizing the possibility of such an effect is a stimulus to look for an excess of quasars close to galaxies (Chapter 2), quasars whose visibility might be enhanced by this kind of intensification. Extensive monitoring of quasar fields in the vicinity of galaxies over periods of years is needed to check for this effect.

The other effect is more subtle. Because of the lensing geometry discussed above, only a small volume of the quasar is amplified by a point mass of stellar size. The internal structure of quasars is such that the size scale of the continuum-emitting region, the broad-line region and the narrow-line region differ greatly (Chapter 9). The narrow-line region of extent $\sim 10^2$ pc would never be amplified in this way, whereas the continuum or broad-line region within 0.01 pc could be. So a population of quasars showing no narrow-line region might be expected were this population all lensed by small objects. Or, if the continuum region is smallest of all, it might undergo amplification without amplification in any emission lines. Unfortunately, such 'continuum flares' are seen in quasars seemingly for completely different reasons (Chapter 9). Because of the other plausible explanations, I am not confident that spectroscopic differences will ever provide believable evidence of quasars lensed by small masses, but the issue of tracking down the dark matter in galactic haloes is so important that all paths must be explored. As a result, looking for systematic spectral differences or variability for quasars proximate to galaxies is a task that should be undertaken.

The point mass approximation is valid for lensing by single stars or stellar remnants. Much worse complications enter for real galaxy-sized lenses in the universe, because these are not point masses. The distributed mass in a cluster of galaxies seems to account for most of the splitting seen in lensed quasars actually observed so far. For the observer, the advantage of cluster lensing is that more mass is involved so the splittings become greater, and the several lensed cases with splittings of order 5″ are explained in this way. Nevertheless, a simple illustration can be given of how one might proceed to predict the number of lensed quasars expected, using the simple case primarily to call attention to the realistic modifications which are necessary.

The first step is to consider the total number of galaxies in the sky, treating them like point masses for lensing purposes, and assigning a representative mass to each galaxy. Also, distance ratios oM/oq have to be assumed; this ratio is taken as 0.5 for all galaxies in this simple case, meaning the lensing galaxy is always assumed to have half the proper distance of the quasar so δ = splitting. Typically, this means $z \approx 1$ for representative lensing galaxies. Using the expression $\delta'' = 1.2 \times 10^{11} \, Mb^{-1}$, the impact parameter b

subtended by a galaxy then determines the area of the sky $\pi\phi^2$ around the galaxy within which a background quasar would be subject to image splitting $\geq\delta''$. The total fraction of the sky so subject is the same as the fraction of quasar images expected to be split by this amount or more. That is, each galaxy produces a 'lens size' of $\pi\phi^2$ in the sky, so the total area of lenses is $N_G\pi\phi^2$. One major convenience in this calculation is that ϕ is a metric angle, $\phi = b(1 + z)/d$, which we already know from Chapter 3 to be virtually independent of proper distance d for $z \gtrsim 1$.

First, we ask how many quasars might have large, easily detectable splittings of $\sim 5''$. Such a value characterizes the first few lensed images actually found. If the characteristic galactic mass is 10^{12} $M_\odot$, $\delta = 5''$ is achieved for $b = 2.4 \times 10^{22}$ cm $= 8$ kpc. The angular size $\phi = b(1 + z)/d \approx 1.2''$ for all relevant z. The density of galaxies on the sky brighter than magnitude m is $N_G \approx (1.5 \times 10^4)10^{0.44(m-24)}$ deg^{-2} (Tyson & Jarvis 1979). How faint should we count galaxies for this illustration? It is appropriate to consider galaxies as faint as they have been seen to make the counts quoted, which is $m \approx 24$, so we take $N_G = 1.5 \times 10^4$ deg^{-2}. For the ϕ deduced, this means that within each deg^2 of sky, there is 6.7×10^4 arcsec^2, or a fraction 5×10^{-3}, covered with intervening galactic lenses capable of producing $\geq 5''$ splittings. This is the same fraction of quasars which should be affected by the lensing. As around one thousand quasars were surveyed in such a way as to reveal splitting to produce the first four lensed cases, this prediction is consistent, within the crudeness of the attempt, with the split images that have been found. It is not, however, a correct explanation of those images.

Note several important implications of even this simple model. The splitting required, δ'', goes as b^{-1}, and the lens area as b^2, so the lens area around each galaxy is as δ^{-2}. This means that the fractional number of split quasar images should decrease with the square of the splitting amount. Many more quasars should be found with small splittings. This explains why it remains curious that the quasar images with small splittings have not been found. The explanation seems to be that the point mass approximation is not valid for splittings, so that the large splittings observed do not arise as in the simple example, but instead come from the distributed mass in clusters.

Because of the many parameters involved in lensing predictions, real models are very complex (Turner, Ostriker & Gott 1984). Correct distributions of galaxies have to be known as a function of mass, position within or without a cluster, and proper distance. Comparison with quasar observations requires determination not only of splitting but also of intensification. The reason is that observed quasar samples are controlled by the magnitude limit of the observation, and intensified quasars thereby gain an artificial place within any unlensed sample. Quasar counts have to be corrected, therefore, for the 'contamination' by lensed quasar images. Fortunately, this contamination seems to be small, but it does happen. The difference in slope of counts as a function of magnitude for quasars compared with that of galaxies (Chapter 2) is evidence that the majority of quasars are not being made visible because of galaxy lensing (Tyson 1983).

The most bizarre image that has been attributed to lensing is that for a high redshift quasar, $z = 1.7$, appearing so close to the center of a spiral galaxy, within 1″, that it was mistaken for the galactic nucleus (Huchra *et al.* 1985). In this case, the disk shaped matter distribution in the galaxy is interpreted to act as a single focusing lens, which re-images and intensifies the quasar by about a factor of 50. Because of the intensification, this image configuration is not improbable, because the surface density of faint quasars is sufficiently high that it is not unlikely to find one occasionally just behind the center of a galaxy. Being boosted by intensification, the intrinsically faint quasar joins the list of much less frequent bright quasars. This quasar image was found as the only case in a survey of 7000 galaxies, so it is not the kind of thing that will appear often.

Available data are now so limited that comparisons between the observed and predicted consequences of quasar images split by gravitational lensing are of marginal utility. Within the broad uncertainties of available data, what is seen is more or less consistent with what would be expected based upon the distribution of known matter in the universe. That is, there is not yet forceful evidence for very large but dark masses being responsible for lensing. Such masses would be required if large splittings were found often without detectable galaxies or clusters as the lens. The greatest difficulty in comparing observations with expectations is that most galaxies should cause splittings smaller than observable in ground based optical surveys, splitting probabilities peaking at about 0.5″ (Turner *et al.* 1984). Until we can test large quasar samples with improved imagery, these expectations cannot be tested.

The demonstration that gravitational lensing is a real phenomenon that actually happens to some quasars unlocked many potential observational tests. Because of all the possibilities of lensing, each of which if tested yields an improved constraint on the distribution of matter in the universe, the quasar observer has an exciting challenge. Large samples are needed to test for lensed pairs on all scales from 0.1″ to 1′. The smaller splittings can only be seen with radio techniques or HST, but ground based optical surveys are sensitive to splittings above 1″. Just looking for close pairs of quasars with seemingly identical spectra is justification enough for continuing quasar surveys. Accumulating samples of quasars near bright galaxies will yield candidates for testing small-object lensing, and determining the spectral properties can provide relevant results without waiting out monitoring time. While many of these observational chores are tedious, the absence of any better way to probe for dark, condensed matter in the universe makes them essential.

5

LUMINOSITY FUNCTIONS

5.1   Fundamentals of a luminosity function

Understanding the true distribution of objects in space has always been a basic objective of astronomy. Highly sophisticated statistical techniques were developed for determining the distribution of stars in our Galaxy (Trumpler & Weaver 1953). Many of these techniques have recently resurfaced for application to quasars. In many respects, there are great similarities between quasar counting, as done today, and the star counting in the early part of this century that led to an understanding of the structure of our Galaxy. Let us hope that similarly significant results may eventually arise from current quasar surveys. It would be convenient to apply the older techniques directly to quasars, just plugging in a few new numbers. This cannot be done, however. Determining the distributions of interest requires dealing with three dimensions, and the cosmological equations that relate distance for quasars to the observable redshift are much more complex than the euclidean geometry usable by galactic astronomers. Furthermore, quasars are not distributed uniformly in the universe, so statistical techniques based upon homogeneous distributions will not work. Finally, all of the equations of statistical stellar astronomy use magnitude units. This is still the case for most optical astronomy of quasars, but not so for quasar counts based on radio or X-ray observations. So we must deal with the additional complication of discussing both magnitude and flux units. It is necessary, therefore, to build a discussion of 'statistical astronomy' for quasars from first principles.

It will never be possible to determine the total number of planets, stars, galaxies or quasars in a given volume of space. The reason is that observations are always flux limited; as a consequence, there will forever remain the possibility that objects exist which are fainter than our ability to see. Such objects could not be included in any accounting of what is there. Because of this fundamental limitation, all descriptions of the number of objects per unit volume, of any category, must be restricted to objects of particular luminosities. The relevant luminosities are those which are visible using the applicable flux limit and relations among flux, distance and luminosity.

Terms used to define the spatial distribution of stars, galaxies or quasars are the 'space density' or the 'luminosity function'. These are not formally the same things, but the terms have sometimes been used interchangeably and without consistency. In what follows, the two terms will be separately defined. It is not the semantics which is important, but the concept of what these terms mean physically. Also, the following discussion will be phrased for quasars, but the techniques apply generally to any category of objects in the universe.

The beginning point is to consider a volume of space, in which is found a number of quasars. Each quasar is slightly different from the others in its intrinsic luminosity. The quasar content of this volume is not necessarily the same as in other volumes of the universe. This proviso is particularly important in the context of quasar evolution considered in Chapter 6. For quasars, therefore, it is necessary to define the volume being considered with respect to the epoch of the universe at which the volume is seen. Quasars within the volume of interest are characterized individually by particular luminosities. Various measurements of luminosity can be used: bolometric luminosities, luminosities for particular spectral regions such as X-ray or optical luminosities, or analogous luminosities defined in terms of absolute magnitudes M.

Each quasar in the volume has a single value for this parameter, and no two quasars have identical values. To describe the number of quasars per unit volume, it is therefore necessary to define the range in luminosity or absolute magnitude within which the number of tabulated quasars are found. Usually, this range is a unit magnitude interval, or a logarithmic interval of luminosity; the intervals have width $\Delta \log L$ or ΔM in luminosity, each interval centered upon values L^* or M^*. The 'space density' of quasars with luminosity L^* is the number of quasars per unit volume having $[\log L^* - (\Delta \log L)/2] < \log L < [\log L^* + (\Delta \log L)/2]$. For magnitude units, a width ΔM of a unit magnitude corresponds to a $\Delta \log L$ of 0.4, because of the definition of magnitude. Analogously, the space density of quasars of absolute magnitude M^* is the number of quasars per unit volume with $M^* + 0.5 < M < M^* - 0.5$.

If the space density of quasars is tabulated for different values of $\log L^*$ or M^*, this tabulation gives the 'space density function'. This tabulates the number in each *interval* of luminosity, which is a differential function.

Common notation for this differential function is $\Phi(\log L^*)$ or $\Phi(M^*)$. To describe quasars in *all* intervals brighter than a given cutoff luminosity, a tabulation that gives all quasars brighter than successively listed values of luminosity should be specifically referred to as an integral function. This integral function is called the 'luminosity function', commonly denoted $\Psi(\log L)$ or $\Psi(M)$. Such a function gives the *total* number of quasars per unit volume brighter than the listed limit $\log L$ or M. Obviously, $\Psi(M)$ is the summation over all bins in Φ containing quasars brighter than M.

5.2   Example of a luminosity function calculation

Many samples of quasars are available with which to calculate luminosity functions or space density functions. Later, several of these samples will be combined. To demonstrate the determination of a luminosity function, and to show how realistic uncertainties in the observations affect the results, the closest large sample of quasars is utilized. As emphasized elsewhere in this volume, nearby quasars are invariably found residing in galactic nuclei. The category of active galaxies whose nuclei are most similar in spectral characteristics to quasars at higher redshifts are the Seyfert 1 galaxies. There are several advantages to making the initial calculation of a luminosity function using a sample of nearby objects. Eventually, it will be necessary to face the complications of volume calculations arising from applicable cosmologies. The important techniques can be demonstrated without that, however, as long as sufficiently nearby objects are considered that the space encompassing them can be approximated as euclidean. Determining the luminosity function of the closest ('local') quasars is an important starting point for investigating quasar evolution, by comparing local quasars with more distant samples. Within the local sample of Seyfert 1 galaxies, evolution with epoch can be ignored. This is justification for proceeding as if we had an ideal sample of objects – one distributed homogeneously.

The optical sample to be used is that of the Seyfert 1 galaxies discovered as part of the Markarian survey for galaxies with unusually blue continua. This survey began appearing in 1967 (Markarian 1967) as an attempt to locate galaxies whose continuous spectra appeared unusually blue. This gave both a spectroscopic and a morphological criterion; objects not only had to be blue but also to show an extended (galactic) image. For this reason, spatially unresolved quasars were not often found. Most of the Markarian galaxies turned out to be characterized by blue stars. But the characteristic continuum shape of quasars also makes them appear blue, which remains the primary detection criterion for optically discovered quasar samples

(Chapter 2). Not surprisingly, therefore, Markarian's criteria proved good ones for also finding Seyfert 1 galaxies. Markarian recognized this, especially in later lists, and called attention to Seyfert candidates.

There have been several other optical surveys whose criteria are such as to locate some Seyfert 1 galaxies. Routine spectroscopic observing, one at a time, of many galaxies reveals occasional Seyfert galaxies (Huchra, Wyatt & Davis 1982). The various objective prism surveys mentioned in Chapter 2 also reveal such galaxies. Looking for galaxies on purely morphological criteria, attempting to find those with intense luminosity concentrations, led to the discovery of some Seyfert galaxies by Zwicky (Sargent 1970) and in a similarly designed survey by Fairall (1984). The primary advantage of the Markarian survey is that it is an all-sky survey, covering $10\,000$ deg^2, carefully described and published, and going sufficiently faint to reveal a large galaxy sample. About 1500 galaxies are now included as Markarian galaxies (see Markarian, Lipovetsky & Stepanian 1981 for the final instalment). Of these, almost 10% were subsequently confirmed as Seyfert 1 galaxies, primarily by Osterbrock and collaborators (Osterbrock & Dahari 1983).

I summarize in Table 5.1 the Markarian galaxies brighter than magnitude 15.5 which qualify as Seyfert 1 galaxies. This criterion as used here only

Table 5.1. *Seyfert 1 galaxies from Markarian survey < 15.5 magnitude[a]*

Markarian number	cz (km s^{-1})	B mag	IRAS flux 25 μm (Jy)	60 μm (Jy)	$\alpha(B - 60)$
6	5670	14.8	0.68	1.11	−1.12
9	11700	15.2	0.53	0.87	−1.15
10	8700	14.0	0.30	0.84	−0.91
42	7200	15.2	—	—	—
50	7020	15.5	—	—	—
141	11700	14.5	—	0.75	−0.99
231	12300	14.1	8.56	33.26	−1.68
279	9150	14.5	0.31	1.08	−1.06
290	9030	15.0	—	—	—
291	10560	15.0	—	—	—
304	19710	14.6	—	—	—
315	11640	14.8	0.41	1.49	−1.18
335	7740	14.0	—	—	—
352	4650	14.8	—	—	—
358	13620	15.0	—	—	—
372	9240	15.0	—	—	—
382	10200	15.5	—	—	—

Table 5.1. *Continued*

Markarian number	$cz\,(\mathrm{km\,s^{-1}})$	B mag	IRAS flux 25 μm (Jy)	60 μm (Jy)	$\alpha(B-60)$
423	9570	14.9	—	1.36	−1.18
471	10260	14.5	—	0.64	−0.95
474	11730	15.3	—	0.52	−1.09
478	23700	15.0	—	0.59	−1.03
486	11700	15.2	—	—	—
493	9450	14.9	0.29	0.64	−1.03
506	12930	15.3	—	—	—
509	10200	14.5	0.75	1.40	−1.11
516	8550	15.4	—	1.35	−1.28
530	8790	14.4	—	0.82	−0.99
541	11850	15.5	—	—	—
584	23640	14.0	—	—	—
590	7890	14.0	—	0.53	−0.82
609	10260	14.5	0.48	2.55	−1.24
618	10860	14.5	0.80	2.73	−1.25
662	16590	15.5	—	—	—
699	10260	15.4	—	—	—
744	2700	13.5	—	—	—
766	3870	13.7	1.38	4.01	−1.18
771	18900	15.1	—	—	—
841	10920	14.0	0.45	0.51	−0.81
883	11400	15.2	—	1.10	−1.20
885	7500	15.0	—	—	—
915	7170	15.0	0.38	0.44	−0.97
975	14730	15.0	0.41	—	—
1018	12720	14.6	—	—	—
1034	10500	14.9	0.70	6.53	−1.50
1040	4920	13.9	1.32	2.73	−1.14
1044	4890	15.5	—	—	—
1095	9900	14.6	0.48	0.65	−0.98
1098	10590	15.5	—	—	—
1126	3090	14.5	—	—	—
1146	11580	15.5	—	—	—
1152	15810	15.0	—	—	—
1187	13470	15.5	—	—	—
1218	8580	14.8	—	—	—
1239	5820	14.5	1.21	1.39	−1.11
1243	10560	14.5	—	—	—
1347	15090	15.4	—	0.90	−1.19
1383	25860	15.0	—	—	—
1400	8790	15.5	—	—	—

[a] B magnitudes are photographically determined Zwicky magnitudes listed by Markarian. IRAS fluxes are from Lonsdale *et al.* (1985).

requires the presence of any broad wings to a Balmer emission line. More subtle spectroscopic classifications are available and are meaningful in terms of the nuclear structure. As discussed in Chapter 9, however, the visible presence of any broad emission line region, no matter how weak, is the most consistent indicator of a quasar. Markarian deliberately searched for galaxies, recognizable as such on his objective prism survey plates. This differs from quasar surveys for which the spectroscopic criteria are similar, but only objects that are starlike (unresolved in angular extent) are accepted. This explains why Markarian found primarily low redshift Seyfert 1 galaxies rather than higher redshift quasars, even though it was the nuclei of these galaxies that actually attracted Markarian's notice, and these nuclei are very similar in appearance to the objects which other surveyors would classify as quasars. I think that most of the Seyfert 1 galaxies within the Markarian survey have now been recognized, as spectra of 70% of all Markarian galaxies are available, and nearly 100% of all which Markarian suspected to be Seyferts. Table 5.1 should, therefore, be reasonably complete as an optically derived sample of Seyfert 1 galaxies. As such, it is a useful list to illustrate the determination of a luminosity function.

The attempt to achieve 'complete samples' is the supreme goal of many who catalog things in the sky. A complete sample is defined as a group of objects with particular, classifiable characteristics found to a precisely described flux limit, with a precisely defined search technique. Despite claims to the contrary, there is no such thing as a complete sample. Understanding the data in hand is where artistry enters astronomy. The purpose of collecting a complete sample would be clear, as the idea is to remove any subjective, preconceived opinions the observer might have as to what will be found. But it cannot be done. No two objects in the sky are exactly alike. There are too many morphological and spectroscopic characteristics available from which to choose. It is not possible, therefore, to compile any catalog of celestial objects that is truly and perfectly complete, without building in the cataloger's opinion of what unwritten characteristics define the sample. It is just as important, probably more so, to understand and accept all the assumptions that enter any astronomical compilation as it is to exercise diligence and statistical sophistication in assembling and analyzing the sample. All samples in astronomy are arbitrary and approximate to some degree. As a result, all deductions concerning luminosity functions will have major associated uncertainties. As long as one can gain a feeling for the level of those uncertainties and the reasons for them, honest astronomy is being done. I think a distinguishing characteristic of a 'good' astronomer is to know when the data are adequate for the assumptions, so that no further effort at the telescope is justified.

In this spirit, we will accept the Markarian sample of Seyfert 1 galaxies, not as perfectly complete, but as having characteristics that are understood within some limits. Where the uncertainties enter the deductions will be pointed out. To start that exercise, the magnitudes for Markarian galaxies are not very accurate. Those listed in Table 5.1 are those given in the Markarian discovery lists, but actually are from old measures in a large scale photographic survey of galaxies by Zwicky. Magnitudes could be measured much more accurately. Is there a point in doing this? Such decisions illustrate the qualitative judgments that must be made. It would be a lot of work to determine accurate magnitudes for all of the galaxies in Table 5.1. Even with the determination to do so, what should be measured? Are we interested in the magnitude of the entire galaxy, or only in that of the nucleus, which is the part that is the quasar? Most of the light comes from the nuclei anyway, so maybe the uncertainty in deciding this fraction outweighs the uncertainty in the Zwicky magnitudes already available. And what about the uncertainties in Markarian's own limits? Every one of his survey photographs was a little different, but by an amount not precisely knowable. Each photographic plate has slightly different sensitivity, and observing conditions vary for each. Maybe he included galaxies from one part of the sky that would have been overlooked on plates from another part. Maybe Seyfert 1 galaxies, within Markarian's range, are not truly homogeneously distributed anyway. Answering each of these questions requires a decision and some assumptions. At some point, one decides to go for an answer, making do with the data that exist. No individual astronomer can completely and rationally detail the series of decisions made to determine when that point was reached. The easiest way out is to lump all of the uncertainties into a single error estimate for the final answer, which is the procedure I follow below.

The objective is to take the data in Table 5.1, assume initially that the magnitudes given are precisely correct, and that the redshifts can be used with a Hubble constant to yield precise distances, and then calculate the luminosity function that results. Afterwards, we can ask how this result might change if some of the assumptions are wrong.

For low redshifts and euclidean space, the luminosity L of any galaxy relates to flux f by $L = 4\pi(cz/H_o)^2(3 \times 10^{24})^2 f$ in the units of Chapter 3. At negligible redshifts, it is irrelevant whether L and f are in bolometric units, $d\lambda^{-1}$ units or $d\nu^{-1}$ units. From the definition of absolute magnitude M, this relates to apparent magnitude B by $M = B - 5 \log (cz/H_o) - 25$. Volume calculations are also simple; $V = (4\pi/3)(cz/H_o)^3 \text{ Mpc}^3$. If it were known what volume of space had to be surveyed to find these galaxies, the calculation of differential space densities Φ would be straightforward: determine the M for each galaxy, bin the galaxies in intervals of $M^* \pm 0.5$, and divide by the volume surveyed. The complication is that surveys are to a given apparent magnitude, or flux limit, and not to the edge of a defined volume of the universe. Galaxies that have bright absolute magnitudes can be seen further into the universe than faint galaxies, so the survey volume applicable to each galaxy depends on its absolute magnitude. To accommodate this selection, the surveyed volume has to be calculated independently for each galaxy using the procedure described by Huchra & Sargent (1973).

Within a survey carried to a limiting apparent magnitude B, it is assumed that any galaxy brighter than B is detected. (Determining the applicable limiting magnitude is the primary challenge of any survey; more about this later.) A galaxy i having luminosity distance d_i in Mpc would be detected only if $B - M_i < 5 \log d_i + 25$, M_i being the absolute magnitude of this individual galaxy. M_i is determined from the galaxy's actually observed magnitude B_i, which is not the same as the limiting survey magnitude B. Setting $B - M_i = 5 \log d_{max} + 25$ gives the maximum luminosity distance at which this galaxy *could* have been detected. The volume corresponding to d_{max} is the true volume that was actually examined to find this galaxy, rather than the volume corresponding to d_i. If the survey covers all 4π steradians, then this galaxy could have been *anywhere* within volume $(4\pi/3)(d_{max})^3$ and been counted. This is the volume occupied by the galaxy of M_i. The inverse of this volume is the space density of galaxies having M_i, but the space density function requires binning galaxies in magnitude intervals. Φ is found by summing the space densities of all galaxies whose M_i fall within the range $M^* \pm 0.5$. That is, $\Phi(M^*) = \Sigma_i \, (4\pi/3)^{-1}(d_{max(i)})^{-3}$. This sum must be increased by the factor $4\pi/\Omega$ if only a solid angle Ω of the sky is included in the survey, to compensate for the portion of the sky in which galaxies were not counted. Some such portion is invariably omitted, if only because the Milky Way obscures the extragalactic background. Usually, surveys are

Table 5.2. *Alternative luminosity functions for Markarian Seyfert 1 galaxies*

M	$\log \Psi(M)$ if limit = 15.0	$\log \Psi(M)$ if limit = 15.5	$\log \Psi(M)$ graphical, limit = 15.5
−18	−4.6	−4.7	−4.8
−19	−4.7	−5.0	−4.9
−20	−5.1	−5.3	−5.0
−21	−5.6	−6.1	−6.0
−22	−6.8	−7.3	−6.8
−23	−8.1	−8.4	—

quoted as covering n deg^2, in which case the factor $4\pi/\Omega$ is $41252/n$. Finally, the integral luminosity function $\Psi(M)$ is the sum over all Φ bins brighter than M.

Following this procedure, the $\Psi(M)$ is calculated for the galaxies in Table 5.1 and tabulated in Table 5.2. That accurate knowledge of the survey limit B is critical to the success of this technique is demonstrated by making alternative calculations using $B = 15.5$ and $B = 15$; the resulting $\Psi(M)$ differ by about a factor of two.

An alternative method to determine $\Psi(M)$ uses only individual redshifts together with the limiting magnitude for the entire sample, avoiding the requirement for individual magnitudes. To illustrate this method, the sample of Seyfert 1 galaxies in Table 5.1 is used again. The redshift distribution of these galaxies is shown in Figure 5.1, arbitrarily taking bins of width 1000 km s^{-1}. The ordinate shows the number of galaxies found per redshift interval of 1000 km s^{-1}. Each bin represents a successive shell in space. The thickness of these shells would be made arbitrarily small by decreasing the bin width; 1000 km s^{-1} is used here to allow a reasonable number of galaxies in each bin. The observer must smooth the data to fit all of the bins with a reasonable curve. Uncertainties in the method are represented by the uncertainties in making this fit.

Once a smoothed fit is adopted, the $\Psi(M)$ can be determined for this sample. This is possible because the number of galaxies in each bin, dN, is given by $dN = \Psi(M)\, dv$. The volume element dv is the volume observed in the shell of space corresponding to any bin. The Markarian survey covered about 10^4 deg^2, one-quarter of the sky, so the volume examined within any shell is $dv = 0.25\, 4\pi(cz/H_o)^2(1000/H_o)$. The velocity cz can be taken as that

Figure 5.1. Histogram: binned redshift distribution in units of 10^3 km s^{-1} for Markarian Seyfert 1 galaxies in Table 5.1. Curve: smoothed fit used to derive 'graphical' luminosity function in Table 5.2.

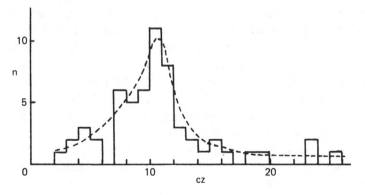

to the center of the bin. Using this, $\Psi(M) = dN\, \pi^{-1}(cz/H_o)^{-2}(1000/H_o)^{-1}$. The M is the absolute magnitude brighter than which galaxies in the bin would be included, or $M = B - 5 \log (cz/H_o) - 25$. Proceeding from bin to bin, the $\Psi(M)$ is determined for progressively brighter values of M. Tabulating the result in Table 5.2 shows good agreement with the other calculation of $\Psi(M)$ for the same B. As before, the major source of uncertainty in the overall result depends on the choice of limiting magnitude B.

5.3   Completeness limits of samples

How can the true limiting magnitude or flux of a survey be determined? It is possible, in principle, to set the survey limit bright enough that the galaxies found are very conspicuous to the search technique used, giving large signals compared with any sources of noise or inconsistency in the search technique. Some galaxies may still be found fainter than the limit claimed, and they must be excluded from the sample. Determining the survey limit in this way essentially requires trust in the ability of the surveyor to find everything to the magnitude limit he claims to reach. As that is often a qualitative or subjective judgment, it is generally considered unsatisfactory to choose survey limits in this way. Most surveys of necessity push the limits of the applicable technique, in order to discover enough objects to compile meaningful luminosity functions. Near such limits, it is exceedingly difficult to evaluate just from the principles of the technique how faint the search really reaches. So other checks are desirable.

These other checks, as now detailed, absolutely require the assumption that the galaxies surveyed are distributed uniformly in space. With that assumption, it is possible to recognize the magnitude at which survey incompleteness begins to set in.

As explained by Trumpler & Weaver (1953) for the same test applied to star counts, the number of objects detected should increase by a factor of four $(10^{0.6})$ for each unit increase in the survey limit B, regardless of the shape of Ψ. If the sample of objects being examined all had the same absolute magnitude M and were above the detection limit for any distance, then the number detected would obviously increase simply with the total volume examined. Because volume goes as d^3 and B increases as $5 \log d$ for constant M, the volume examined increases with $10^{0.6\Delta B}$. This applies if the object distribution is uniform and of unlimited extent, and if space is perfectly transparent, all within euclidean geometry for space. These assumptions are so restrictive that it is not realistic to apply them for most quasars. For nearby galaxies, it is reasonable to make this check as a crude test of limiting magnitude.

The reason that this numerical result also arises for a distribution of M is because the spatial distribution is taken as unlimited. Observing at any magnitude B gives access to all M, somewhere in space, with the M being accessed depending on the distance to the galaxy. If B increases, the distance to which a given M can be detected also increases. The volume of the shell containing objects within any one magnitude interval $M \pm 0.5$ increases in proportion to d^3, as the distance at which that particular M is found increases with B. As long as the Ψ has the same *shape* throughout space, the number of objects seen with any M will therefore increase with $10^{0.6\Delta B}$, so the totality of objects in Ψ must also increase by this factor.

The test for incompleteness in a magnitude limited survey, therefore, is to plot the number of objects found as a function of magnitude. That magnitude at which the number ceases to increase at the expected rate $10^{0.6\Delta B}$ is an estimate of the survey limit. Using this technique, the limit of the Markarian survey has been found to be between 15 and 16 mag. Completely analogous to this test is the notorious 'log N – log S' plot used early in the history of quasar astronomy to check the distribution of radio quasars in the universe. With the same set of assumptions, N being the number of sources detected and S now a flux limit rather than a magnitude limit, it can be seen that $\log N \propto 1.5 \log S$. This is because $\log N \propto 0.6 \Delta B$, but $B \propto -2.5 \log S$, for the transformation between flux and magnitude units.

I think that the test described has often been misused as a determination of a survey limit. The reason is the requirement that objects considered should be unlimited in extent and maintain the same Ψ throughout the observable volume. This is not so for any galaxy or quasar survey. All such surveys have built-in redshift limits, so objects found are constrained to certain distances. Consider the extreme situation, which does arise in some quasar surveys, where the survey is directed to a well defined interval of redshift. It will then make no difference how faint the limit is pushed; quasars will not be seen at greater distances, but only fainter quasars within the defined redshift interval will be found. In that case, going to fainter magnitudes simply describes the shape of Ψ within this redshift interval. That is a very useful thing to do, but when doing so it is meaningless to apply the test described above for survey completeness. Analogous objections, though not as severe, apply to galaxy surveys, such as the Markarian survey. That survey, as previously mentioned, is constrained by the requirement that objects found should be recognizable as galaxies. Accordingly, such objects are restricted to distances corresponding to $z \lesssim 0.1$, above which galaxies may not be recognizable as extended objects on plates from survey telescopes. Even though galaxies at greater distances might have M that fall within the survey's reach, they would not be included if not recognized as galaxies. So the N vs. m plot would begin to fall below the rate of increase expected, even if the limiting magnitude had not been reached.

There is an alternative test for limiting magnitude, subject to the same assumptions already described. It is useful to describe this test because its greatest utility has come in checking the assumption of uniform distribution for quasars, rather than in determining limiting B. This is the 'V/V_{max}' test, discussed more thoroughly as applied to quasars in Chapter 6. The idea is that for a random but homogeneous distribution of galaxies, a galaxy of given M_i is just as likely to be found within the inner 50% of the volume surveyed as within the outer 50%. The actual distance to the galaxy is $d_i = cz_i/H_0$, so this galaxy has been found within the volume $V_i = 4\pi d_i^3/3$. Yet, it could have been found anywhere within the distance d_{max} (determined from survey limit B and galaxy magnitude M_i, as in Section 5.2) or within the volume $V_{max} = 4\pi d_{max}^3/3$. For the entire sample of galaxies, one would expect that half should be found with $V_i < 0.5\ V_{max}$ (in the inner half of the surveyed volume), and half with $V_i > 0.5\ V_{max}$ (in the outer half of the surveyed volume). The mean V_i/V_{max} for the entire sample must, therefore, be 0.5, if the galaxies found are distributed uniformly. If the survey is becoming incomplete at some magnitude B, galaxies will preferentially be found within the inner half of the V_{max}, it being easier to see closer galaxies than distant ones at the marginal detection limits of a survey. The check for limiting survey magnitude, therefore, is to assume a limit, calculate mean V_i/V_{max} for all galaxies found to that limit, and see if this mean is 0.5. If the answer is <0.5, the limiting B assumed is too faint, and calculations of V_i/V_{max} must be repeated for various B until a B is found that is bright enough to yield mean $V_i/V_{max} = 0.5$. Note that choosing B too bright does not make $V_i/V_{max} > 0.5$! There is no way to increase the probability of finding a galaxy in the outer half of the volume surveyed. When mean $V_i/V_{max} > 0.5$ are found, that is a demonstration that the objects are not distributed uniformly, their space density increasing with increasing distance. The great utility of the V/V_{max} test for quasars, discussed later, is to show that this does indeed happen, providing the first proof of quasar evolution in the universe. With enough data, this test can be applied to subsets of a sample to determine if the limiting magnitude depends on various characteristics such as morphological or spectral types. Second-order refinements to survey completeness can be achieved in this way.

For nearby galaxies and quasars, serious impediments to determining Ψ arise from spatial inhomogeneities in the distribution. Gross inhomogeneities are known to exist within $z \lesssim 0.1$ for some galaxies (Kirshner *et al.* 1981, Tarenghi *et al.* 1979). Clever techniques have been developed to determine luminosity functions under such circumstances (Kirshner, Oemler & Schechter 1979), although the ability to check the survey limit independently is lost. For want of enough data to measure such inhomogeneities, all of the local quasar luminosity functions described herein assume a homogeneous spatial distribution for $z \lesssim 0.1$. This is a separate issue from systematic changes in Ψ as a function of z, which is the problem of quasar evolution discussed in Chapter 6.

Regardless of the quality of the observations which are made to collect data such as those used to produce this result, there is a fundamental limitation to precision set by an uncontrollable factor. Hidden in the assumptions mentioned above was the statement that space had to be transparent, meaning that nothing interfered with light in its passage to us. Unfortunately, something does interfere. For optical observations, it is the dust associated with the disk of our Galaxy, so apparent magnitudes of galaxies and quasars are made fainter by this 'galactic extinction'. Heavy elements associated with the interstellar gas in the disk absorb X-rays, particularly those with energies less than 1 keV. Corrections are made to attempt to account for these effects. Optical magnitudes of galaxies are usually corrected in proportion to the path length through the disk, which means the corrections go as the cosecant of galactic latitude. There is disagreement what the scale factor should be; one school holds that there is no galactic extinction when looking nearly perpendicular to our disk (galactic latitude $= b = 90°$). Another feels that even there extinction may amount to ~0.2 mag. Unquestionably, extinction is different for different objects. Probably it is different in a very complicated way. Observations of the far infrared sky with the Infrared Astronomical Satellite (IRAS) revealed interstellar dust via its infrared emission, and showed the sky to be covered with irregular 'infrared cirrus' (Low *et al.* 1984). Dust in the galactic disk is not uniformly distributed, so corrections based on cosec b are very approximate; maybe even worse than no correction at all. The reason is that corrections are inconsistent, some use one form and others another, so a user of the data must know how to de-correct it before seeing what the observations really were. As long as galactic latitudes are above about 30°, extinction corrections do not exceed a few tenths of a magnitude, and if observations are above about 60°, the differential corrections from object to object are less than 0.1 mag. I prefer to accept this as an unknowable source of uncertainty, rather than to apply an arbitrary correction. The issue becomes somewhat serious in comparing quasar counts in different parts of the sky, because (as seen in Chapter 2) such counts are very sensitive to the magnitude limit. The best advice is to stick to high galactic latitudes when studying extragalactic objects, and hope that uncertainties introduced by galactic extinction are not significant compared with other uncertainties of observation and assumption. Corrections to X-ray observations are even more arbitrary than to the optical, usually correcting for the same amount of absorption regardless of direction.

I hope the lesson learned from the preceding discussion is that it is never possible to be highly precise in describing the limiting magnitude, or flux, of a survey. Many assumptions define the nature of the survey, and there are no truly independent and satisfactory checks on the completeness that do not also require assumptions of homogeneous spatial distribution. Such is the basis for my earlier statement that there can never be a truly 'complete sample'. For the example considered, Seyfert 1 galaxies from the Markarian survey, we can be reasonably assured that all of the uncertainties are incorporated if a range of 15 to 15.5 mag is allowed for the completion limit; the resulting values of Ψ (Table 5.2) are the limits of uncertainty that bracket the real value. Applying different incompleteness corrections explains why different Ψ appear in the literature based upon the same data set.

5.4   The local luminosity function for quasars

The analyses presented above of Markarian Seyfert 1 galaxies provide a start toward determining the local quasar luminosity function. Other samples are available to check the results. Before proceeding to them, it is important to emphasize that the objective of determining a quasar luminosity function means that the results must be restricted to the luminosity of the nuclei of Seyfert 1 galaxies, not the entire galaxy. We wish to see the shape of the Ψ for quasars alone, uncontaminated by the stellar luminosity of the host galaxy. For bright quasars, including the brighter nuclei in the Markarian sample, this is no problem because the quasar greatly outshines the rest of the galaxy. This is the reason, however, why the luminosity function using Markarian galaxies does not continue fainter than $M > -19$. By that M, most of the measured luminosity comes from the galaxy of stars, not the nucleus. Not only does this distort Ψ results, but if the nucleus is faint relative to the galaxy, it would not even be found using Markarian's technique. It will be necessary, therefore, to have alternative methods for finding and measuring the less luminous nuclei. Even with the Markarian sample, an important correction can be made using magnitude measurements corrected for the light of the surrounding galaxy, so that the measured Ψ better represents that light just from the nuclei (Fu-Zhen *et al.* 1985).

At the high luminosity extreme, the Markarian sample may miss quasars, because the surrounding galaxy might be too faint in contrast to the bright nucleus. This selection effect can be overcome by using samples of quasars found by spectroscopic or color criteria alone, without regard to the presence of an enveloping galaxy. The most thorough sample for this use is the Palomar Bright Quasar Survey (PBQS) (Schmidt & Green 1983), which found 33 quasars with redshifts less than 0.1 and rediscovered 16 of the Markarian Seyfert 1 galaxies. Using the same precepts discussed in Section 5.2, this sample is used to derive the luminosity function displayed in Figure 5.2. As would be expected from the selection criteria, the PBQS agrees well with the Markarian results at the bright end of the function where

Figure 5.2. Local quasar luminosity function. M is absolute blue magnitude. Ψ has units of number Mpc^{-3} brighter than M, both for $H_o = 75$ km s^{-1} Mpc^{-1}. Arrows: upper limits from survey of Keel (1983). Open circles: quasars with $z < 0.1$ from PBQS. Crosses: Markarian Seyfert 1 galaxies for limit 15.5 from Table 5.2. Vertical bars: range of results in Fu-Zhen *et al.* (1985) for nuclei of Markarian galaxies. m: results of Maccacaro *et al.* (1984) from 2 keV X-ray luminosity function. p: results of Piccinotti *et al.* (1982) from 10 keV X-ray luminosity function.

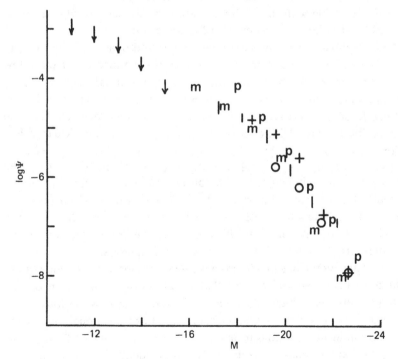

the Markarian galaxy magnitudes are dominated by the nuclei. But the PBQS results show fewer quasars at the fainter magnitudes where the nucleus would not dominate the galaxy, thereby making the objects less likely to be discovered in a survey designed only to find unresolved quasars. This example illustrates how both morphological and spectroscopic criteria enter into quasar surveys implicitly, even though the presence of such morphological selection is not always carefully described. It explains, for example, why 'quasar' surveys and 'Seyfert galaxy' surveys may not precisely overlap at low redshifts, even though I contend that the two categories are the same.

Probing the faint end of the quasar luminosity function is a major challenge, even for nearby quasars. We know that faint galactic nuclei exist that have all of the spectroscopic and variability characteristics of quasars (Chapter 9). The difficulty is in finding a luminosity measurer that allows the quasar luminosity to be determined independently of the contaminating galaxy starlight. With techniques presently available, the best such indicator seems to be the broad line component of the Hα emission line. This is unambiguously associated with the quasar's activity and is reasonably easy to measure, even in faint nuclei. Several efforts have shown that 'mini-quasars' detectable with this technique are in many nearby galactic nuclei (Keel 1983, Stauffer 1982, Filippenko & Sargent 1985).

Enough galaxies have not as yet been measured quantitatively to weak spectroscopic limits to determine a firm luminosity function for these low luminosity quasars. The best results only yield upper limits by accounting for even fainter quasars that might be present but not yet detected. I have modified Keel's results to deduce such limits by taking any detected Hα as an upper limit to the broad line component, and applying an empirical transformation to M using results for other Seyfert 1 galaxies. This is $M = -2.5 \log [L(H\alpha)] + 84.70$, where $L(H\alpha)$ is the Hα luminosity in erg s^{-1}. The resulting Ψ is shown in Figure 5.2; obviously better data to change the limits into real values are highly desirable. It would be very nice to have an entire quasar luminosity function, bright and faint, based only upon the broad Hα line. There is no impediment to this other than the present lack of quantitative measures for enough Seyfert 1 galaxies and quasars.

There is always the nagging fear that some very important selection effect might have been overlooked and that results derived only from optical techniques may give highly erroneous answers. The most serious reservations arise from the worry that quasars might be dusty enough to obscure the visible indicators of quasar activity (Rieke & Lebofsky 1979). Fortunately, we have available a completely independent backup to the local luminosity function calculation. This arises from surveys for quasars based upon their X-ray flux. There have been determinations of the local X-ray quasar luminosity function with two X-ray experiments, at both 2 and 10 keV (Maccacaro, Gioia & Stocke 1984, Piccinotti *et al.* 1982). The comforting result is that such surveys find the same Seyfert 1 galaxies as revealed by the optical surveys. The more difficult part comes in attempting a quantitative normalization of the optical and X-ray results. That is, what optical magnitude M corresponds to a 2 or 10 keV X-ray luminosity?

Determining the appropriate conversion is complicated by many things, the most serious being the lack of knowledge concerning the X-ray spectra. The X-ray detectors are not monochromatic, but detect total X-ray flux over a wide energy range with varying efficiency. This means that extracting a monochromatic flux requires applying the detector efficiency curve to an assumed spectral shape, folded in with an assumed value for absorption by intervening gas. The Einstein IPC, for example, yields detections quoted as 2 keV measures which actually arise from the integrated flux in the 0.3–3.5 keV energy bandpass. Most quasar measures in the X-ray have been made with this detector, and parameters have been defined to relate the results to other parts of the spectrum. The most common is the parameter α_{ox}, defined as the equivalent power law index needed to connect the 2 keV flux to the flux at 2500 Å (Tananbaum *et al.* 1979). Formally, $\alpha_{ox} = -0.38 \log [L(2 \text{ keV})/L(2500 \text{ Å})]$. From X-ray observations of optically selected quasars, this parameter has a mean value of 1.46 (Zamorani *et al.* 1981). From optical observations of X-ray discovered quasars, the mean is 1.26 (Gioia *et al.* 1984). I will use the average of 1.36 for further calculations.

Adopting the standard spectral index of -0.5 used herein for the optical continuous spectrum (Chapter 7), the relation between apparent magnitude B and 2 keV flux is $B = -2.5 \log f(2 \text{ keV}) - 57.57$. Between absolute blue magnitude M and 2 keV X-ray luminosity, it is $M = -2.5 \log L(2 \text{ keV}) + 42.6$. These transformations allow interchange between a luminosity function determined from 2 keV luminosities and one based on absolute magnitude M.

Incorporating the higher energy results of Piccinotti *et al.* (1982) is even more uncertain because there are less data to normalize between the X-ray and optical fluxes. Nevertheless, this survey is important because it covered a significant portion of the entire sky. Of the 85 sources found at high galactic latitudes, 23 are Seyfert galaxies for which an X-ray luminosity function was derived. This is expressed in terms of the total luminosities in the 2–10 keV band, rather than a monochromatic luminosity. The $L(2$–$10 \text{ keV})$ defined arises from assuming an X-ray continuum of shape $\nu^{-0.7}$, folded through the detector bandpass. The bandpass is stated to extend from 3 to 17 keV, within which response exceeds half of the peak quantum efficiency. Integrating the assumed continuum over this bandpass gives $L(2-10 \text{ keV})$, which could be related with these assumptions to a monochromatic luminosity. Choosing 2 keV, the $L(2 \text{ keV}) = 7.7 \times 10^{-19} L(2-10 \text{ keV})$ for $L(2 \text{ keV})$ in erg s^{-1} Hz^{-1} and $L(2-10 \text{ keV})$ in erg s^{-1}. This yields the transformation between the total 2–10 keV luminosity and the absolute magnitude; $M = -2.5 \log [L(2-10 \text{ keV})] + 87.9$. This is the transformation used to place the results of Piccinotti *et al.* on Figure 5.2, simply by integrating their differential function.

A general caution is necessary when relating different luminosity functions. They are often derived with different assumed values of H_o. All values expressed here use $H_o = 75$ km s^{-1} Mpc^{-1}. To transform results for luminosity functions derived with other H_o into results equivalent to those here, the other luminosity scale must be multiplied by a factor $(H_o/75)^2$, or the other absolute magnitudes changed by an additive amount equal to 5 log $(75/H_o)$. Similarly, all space densities from other functions must be multiplied by the factor $(75/H_o)^3$. The reasons for these transformations are straightforward; if H_o less than 75 was used, it would imply a more extended universe, making calculated luminosities greater but thinning out the galaxies, making calculated densities less.

Now that luminosity functions from various sources have been combined, we can look at Figure 5.2 and see the pleasing result that the functions agree reasonably well. The scatter among them is a good representation of the uncertainties arising from the various assumptions that had to be made. Still, we have quite a meaningful answer that provides a fundamental benchmark for quasars – their number as a function of luminosity in the local universe. For convenience, a single fit to all of the points in Figure 5.2 is tabulated in Table 5.3, for use in evolution calculations in Chapter 6. The proper uncertainty estimate to apply to this tabulation has to be judged from the scatter in Figure 5.2.

Often, it has been assumed that luminosity functions are power laws; this eases quantitative fits to sparse data. That is, it is taken that $\Psi(L) \propto L^{-n}$. For such a fit, a plot like Figure 5.2 would show a straight line whose slope depends on the index n. A power law function could fit the brighter data points in the figure; changing L by a factor of ten (2.5 mag) changes log Ψ by 2.5, so $n = 2.5$. This agrees with slopes typically found (Kriss & Canizares 1982).

An important motivation for determining any luminosity function is to relate the population of objects being studied to other populations. For nearby quasars, the most interesting comparison is with the number of nearby galaxies. This cannot be done simply by comparing quasar magnitudes with galaxy magnitudes, because faint quasars are embedded within bright galaxies. It would be the total galaxy magnitude that is required for comparison with any other population of galaxies. What are needed for an accurate comparison with other galaxies are the magnitudes of the host galaxies, after subtracting the quasar luminosity. The reverse was required when attempting to determine the quasar luminosity function. As of now, sufficient data are not available to compile a luminosity function for the host galaxies of quasars. The seriousness of this issue for the present discussion arises because we must know at what faint limit to begin counting galaxies which can host quasars. Otherwise, the important proportion of galaxies which could potentially harbor quasars, but do not, cannot be measured.

The luminosity function of galaxies increases rapidly in number as absolute magnitudes become fainter; most galaxies in the universe are faint irregulars and dwarf ellipticals. None of these have ever been observed to contain a quasar. The quasars within the local quasar luminosity function are found within the nuclei of full sized spiral galaxies (Adams 1977, Yee 1983). It would be inappropriate to count the dwarf galaxies as a comparison sample to these. In the absence of accurate information on magnitudes of the host galaxies, the only option is to adopt an absolute magnitude limit that is reasonable for spiral galaxies, and then compare the number of quasars with the number of galaxies brighter than that limit. Several galaxy luminosity functions are available for such a comparison (Felten 1977, Kirshner *et al.*

Table 5.3. *Local quasar luminosity function*

Absolute magnitude M	$\log \Psi(M)$
−24	−9.5
−23	−8.4
−22	−7.4
−21	−6.4
−20	−5.5
−19	−5.0
−18	−4.7
−17	−4.4
−16	−4.1
−15	−3.9
−14	−3.7
−13	−3.5
−12	−3.3

1979). For the purposes herein, the uncertainty is more in the choice of the absolute magnitude cutoff than in the particular luminosity function chosen.

All galaxies with $M < -20$ should certainly be included from the galaxy luminosity functions. Brighter than this limit, normalized to $H_o = 75$ km s^{-1} Mpc^{-1}, the determinations referenced above give about 5×10^{-3} galaxies Mpc^{-3}. Were the limit for comparison extended to $M < -18$, the number of galaxies would increase to about 2×10^{-2} Mpc^{-3}. The spread of these numbers by a factor of four represents the uncertainty faced when we attempt to describe the fraction of galaxies which contain quasars. A quantitative result is most important when interpreting quasar evolution in Chapter 6. For that purpose, the result adopted from the numbers above is 10^{-2} galaxies Mpc^{-3}, which are otherwise similar to the galaxies that contain quasars. Looking at the local quasar luminosity function in Figure 5.2, there are at least 6×10^{-5} quasars Mpc^{-3} (taking only the measured points to $M < -16$). If the upper limits are close to the actual values, there may be as many as 10^{-3} quasars Mpc^{-3}. In summary, it appears that 0.6 to 10% of galaxies capable of harboring quasars actually do so.

5.5   Luminosity function at high redshift

To proceed from the nearby domain, where most quasars are recognizable as events in the nuclei of galaxies, to the distant limits of the universe, where only the unresolved blaze of the quasar is seen, is the next objective. The issue of the luminosity function for distant quasars is confusing, because this function is not the same with increasing redshift. Every quasar sample extending to redshifts of order unity has $V/V_{max} > 0.5$, indicating an increasing number of quasars at higher redshifts compared with the expectations of a uniform distribution. The luminosity function is evolving, and disentangling the form of the function from the form of its evolution is the nature of the difficulty faced. The evolution characteristics are discussed in Chapter 7. It is useful now to determine the luminosity function for a set of quasars at one well defined epoch at high redshift, for comparison with the local function.

This exercise is useful for showing the increased complexities which come about when cosmological equations have to be used, as opposed to the simple euclidean relations that were applicable for the determination of luminosity functions in nearby space. The epoch to be considered now is that one at high redshift for which nature has most favored the observer. There is one particularly well explored window into the distant universe where quasars can be found relatively easily to faint magnitudes. This is the window within which the Ly α emission line of hydrogen, the strongest feature in the quasar spectrum, is redshifted into the visible. As discussed in Chapter 2, quasar surveys based on detection of this feature are already available to 22 mag; HST should have the capability to extend such detections to 25 mag. Depending on the size of the sample desired, an arbitrarily narrow redshift window can be examined, sufficiently small in principle to represent a narrow shell, or well defined epoch, in the universe.

In practice, the best compromise between a reasonably narrow redshift window and the need for sufficient quasars for an adequate analysis recommends use of the interval $2.0 < z < 2.5$. Within this interval, data are available to show the luminosity function over a range of six magnitudes (Weedman 1985). Examining that luminosity function can give a simple first comparison with the luminosity function derived for local quasars. It is a comparison extending over an extraordinary time in the universe. The look-back time to the window concerned exceeds two-thirds of the age of the universe (equation 3.19), so it is possible to examine the quasar phenomenon as it was about 10^{10} years ago, and compare it with similar phenomena today. This will be a simple and direct way to begin considering quasar evolution in Chapter 6.

What is needed to begin is the apparent magnitude distribution of quasars in the interval used, shown in Figure 5.3. Two cosmological relations will be needed. One is that yielded by equations 3.16a and b to transform the apparent magnitudes into absolute magnitudes normalized to the same wavelength as quasars observed locally. The second is that yielded by equations 3.21a and b to determine the co-moving volume within the interval $2.0 < z < 2.5$. Both transformations are done for the choice of cosmology $q_0 = 0.1$, which yields the numerical results in Table 5.4. The

Table 5.4. *Quasar luminosity function for* $2 < z < 2.5$

Absolute magnitude M	$\log \Psi(M)$
−29.5	−10.25 ± 0.25
−28.5	−9.3 ± 0.4
−28.0	−8.5 ± 0.2
−27.5	−8.2 ± 0.2
−26.5	−7.2 ± 0.1
−26.0	−6.7 ± 0.1
−25.5	−6.2 ± 0.15
−25.0	−5.8 ± 0.2
−24.5	−5.7 ± 0.25
−24.0	−5.6 ± 0.25

tabulation does not show differences that come about in luminosity functions caused by alternative cosmological assumptions; such results do not help choose the cosmology. For comparison with the local quasar function, much more cosmologically independent, I adopt the luminosity function in Table 5.4 for $q_0 = 0.1$. This function, together with the error bars arising from the uncertainties in the apparent magnitude distribution of Figure 5.3, is reproduced in Figure 6.2 for a direct comparison to the local function. It is from that comparison that the discussion of quasar evolution begins.

Figure 5.3. Apparent magnitude distribution for quasars with $2 < z < 2.5$. B is apparent blue magnitude; n has units of number deg^{-2} brighter than B.

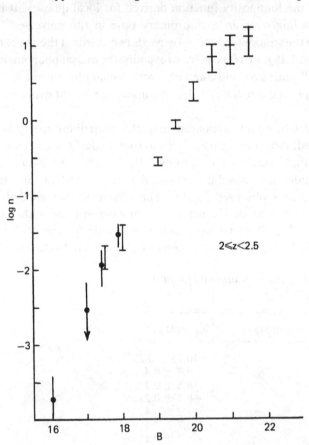

6

QUASAR EVOLUTION

6.1   Basic test for evolution

Determining whether the properties of the universe have changed as a function of its age is a major concern of observational cosmology. Not without logic, a universe maintaining the same characteristics through all of time has a satisfying nature. If we could understand it now, we would by definition understand it always. Even those who do not adhere to such a steady state universe have been loath to invoke changing characteristics to the observable galaxies in the universe. Another one of the ironies of the history of astronomy is that the cosmological tests utilized to prove that we inhabit an evolving Friedmann universe, tests applied using the bright elliptical galaxies as distance probes for cosmological purposes, could not allow evolution of those same galaxies (Sandage 1961). Constancy of the galaxy properties was a necessary prerequisite to using them for cosmological purposes. It is presumed now that such galaxies do change, even over observable time scales, and our ignorance about the proper evolutionary corrections to apply has removed much of the stimulus for drawing cosmological conclusions (Tinsley 1977).

Yet, astronomers who two decades ago accepted little evolution for galaxies were never hesitant to accept a lot of evolution for quasars. Even now, it is necessary to invoke far more evolution in quasars than in galaxies to explain the data seen. Few are troubled by this inconsistency, but it is not too surprising that some are. If we interpret quasar observations in the context of a conventional Friedmann cosmology, either open or closed, dramatic evolution of quasars cannot be avoided. As will be shown, it is necessary to invoke a systematic luminosity decay such that a typical quasar ten billion years ago was 100 times more luminous than a quasar today. No comparable evolution is required for galaxies over the same time frame. So it is fair to worry just a little that something is wrong with our cosmological perceptions. It is a very similar degree of worry as was cited in Chapter 3 on cosmology. An open Friedmann universe comes close to explaining every cosmological test that can be made, except that nagging inconsistency between some values claimed for the Hubble time and the ages of globular clusters. All of the data are not totally self-consistent, so the window of vulnerability still has a little crack. Assuming this will eventually be sealed, I will proceed in this chapter to describe the required form of quasar evolution by utilizing the cosmological equations for $q_o = 0.1$, taken at present as our best-guess universe.

The first step in the development of an evolutionary scenario for quasars was the simple demonstration that they must evolve (Schmidt 1968). This was done with the V/V_{max} test described in Chapter 5. To recap, the principle is that if a sample of objects is homogeneously distributed in the universe around us, and that sample can be observed to a maximum distance corresponding to a volume V_{max}, then half of the objects must be within the inner half of V_{max} and half in the outer half. That is, the ratio of V, the volume examined to find an object, to V_{max} must average 0.5. If it is less, the objects have a greater density nearby; if more, a greater density further away. Simply calculating V/V_{max} for any quasar sample thereby determines if they are distributed evenly or not. This test can be done with any cosmology. Details of the evolution inferred will be cosmology dependent, but the simple question of whether quasars evolve or not can be answered. For a working example, this test is applied to the sample of quasars in Table 6.1, from Marshall *et al.* (1983*b*).

The sample limiting magnitude is 18.3, which determines the z_{max} at which a quasar of the listed magnitude could have been and still have been seen. The z_{max} is found by using equation 3.16*a* (assuming $\alpha = -0.5$) to determine M for the magnitude B and redshift z listed for each quasar. Then, the same equation gives the z_{max} at which this quasar would have $B = 18.3$. Numerically integrating dv in equation 3.21*a* yields both the V and V_{max} for each quasar in Table 6.1. That V/V_{max} averages greater than 0.5 is evidence that, with these cosmological equations, the quasars found were systematically more common at higher redshifts.

At least one meaningful cosmological test can be made using quasars and a definitive answer found. This is a test of any cosmological model which disallows evolution for constituents of the universe. Quasars prove, for example, that the universe is not described by the equations of steady state cosmology. That V/V_{max} for quasar samples is found to exceed 0.5 in the steady state cosmology is the required proof that this is invalid. The steady state cosmology requires that there can be no evolution of anything in the

universe. Any demonstration of evolution for quasars, even with a simple
test like this, rules out the cosmology. Other non-evolving cosmologies can
be similarly tested.

6.2   Parameterizing evolution

Deciphering the way in which quasars have evolved is far more than
an exercise in statistics or semantics. Very fundamental astrophysical questions are at issue. Basic questions are whether quasars have at some time
occurred in all large galaxies and whether most galaxies today contain dead
quasars. Furthermore, it cannot ever be said that the quasar phenomenon is
understood until there is a physical explanation of why quasar characteristics
change with time. When did quasars first form? None of these various
questions can be addressed without an observationally derived description
of what really happens to quasars as a function of epoch in the universe.

Table 6.1. *Faint quasar sample* $(B < 18.3)^a$

Quasar $(AB$ no.)	B	z	$-M$	$V (10^4$ Mpc^3 deg$^{-2})$	z_{max}	$V_{max}(10^4$ Mpc^3 deg$^{-2})$	V/V_{max}
4	17.92	1.27	26.06	350	1.47	470	0.74
7	18.08	2.07	27.15	890	2.24	1020	0.87
9	17.20	1.24	26.72	340	1.90	760	0.45
11	18.22	0.27	22.06	9.4	0.28	10.3	0.91
29	17.81	1.43	26.47	440	1.73	640	0.69
47	17.95	0.23	21.97	6.3	0.27	9.4	0.67
62	17.93	0.27	22.35	9.4	0.32	14.5	0.65
67	18.15	1.50	26.25	490	1.59	550	0.89
78	17.17	1.38	27.02	410	2.15	950	0.43
84	17.99	0.69	24.49	100	0.79	130	0.77
87	17.80	0.77	24.94	125	0.94	190	0.66
89	18.03	1.19	25.79	310	1.33	380	0.82
91	17.38	0.095	20.56	0.5	0.144	2.3	0.22
122	17.84	0.49	23.82	45	0.60	70	0.64
125	18.23	0.28	22.14	10.3	0.29	11.3	0.91
133	17.06	0.18	22.31	3.3	0.31	13.4	0.25
141	17.92	0.94	25.31	190	1.10	260	0.73
142	17.87	0.30	22.66	12.3	0.36	19.5	0.63
147	17.89	0.19	21.60	3.8	0.23	6.3	0.60
154	17.74	2.08	27.51	900	2.59	1320	0.68
162	17.88	1.76	26.93	660	2.06	880	0.75
163	17.54	1.27	26.44	350	1.71	630	0.56
						Mean =	0.66

^{a} From Marshall *et al.* 1983*b*.

For the remainder of this chapter, it is assumed that quasars evolve within a cosmology with $q_o = 0.1$, obeying the equations of Chapter 3. The chore that remains is to determine the form of this evolution that is described by existing quasar data. A physical theory to explain the evolution cannot be sought until the evolution is described. We must understand how the detectability of quasars changes as a function of magnitude and redshift, then transform this understanding into a description of change in the quasar luminosity function with epoch in the cosmology adopted. Whether or not evolution is required for quasars is a test of cosmology only within the philosophical grounds of whether the universe is unchanging or not. Real tests of cosmology should avoid such matters of taste. The adoption of a Friedmann cosmology within which to analyze quasar evolution is a choice based upon common usage of that cosmology for analyzing galaxy observations, subject to the reservations in Chapter 3. Quasars have not, as yet, provided independent confirmation of the choice of a cosmology.

It must be made clear from the start that the entire discussion of evolution to follow cannot be applied to the evolution of a single quasar. The available data simply cannot address that. We can only observe groups of quasars at different epochs in the universe and ask if the properties of these groups change. Then, in a statistical sense, deductions can be made about the changes that would be necessary for individual quasars, but such changes must be on time scales exceeding 10^8 years, so cannot be followed observationally. This situation is directly analogous to that applying to the evolution of stars, using star clusters of different ages to determine how properties of groups of stars change with time, from which the changes in an individual star like the Sun can be inferred even if never followed by observations.

Our understanding of the evolution of quasars is not yet to the point where it is sufficiently precise to specify the physical mechanisms responsible for it. That is the reason why it is so important to pin down the nature of the evolution. Our understanding of what is really happening inside quasars can be greatly improved if we could understand how these events change with time. Several plausible scenarios have been suggested to explain how quasar evolution could occur, but these will not be reviewed until the observational status of the problem is described.

Recalling the discussion of luminosity functions in Chapter 5, it is possible to describe the number of quasars per unit volume as a function of their luminosity. As described in that chapter, the preferred way to do this is simply to list the total number of quasars per unit volume brighter than luminosity L, or $\Psi(L)$. Within the local universe, there are no further modifications. Looking to high redshifts, however, the concept of unit volume changes regardless of whether there is real evolution of the luminosity function. It is necessary to normalize all physical properties, such as density, to a universe in which an observer's volume represents the same fraction of the universe at any epoch. That is, the dilution caused by expansion of the universe has to be taken out before dilutions or enhancements caused by any other effect can be compared. This produces the need for the idea of 'co-moving' density in Chapter 3 which, within an expanding universe, corrects the volume equation by a factor $(1 + z)^3$. This correction is already in equations 3.21a–c. If calculations were made for non-expanding universes, there is no correction by this factor, and the idea of a co-moving volume does not enter.

In order to understand how an evolving luminosity function (one whose nature changes with epoch) affects observed results for quasars, consider a quasar survey to a given magnitude limit. The observer would like to allow for the possibility that the quasar luminosity function changes as some function of redshift z. With this possibility, this observer wishes to subdivide his data into results for different shells of the universe; that is, for different successive redshift increments dz. Then d$N(z)$ is the number of quasars deg^{-2} in the sky (or over any other solid angle desired) within a range dz of redshift centered on redshift z. What determines d$N(z)$? This is set simply by d$N(z) = \Psi[L(z)]$d$v(z)$dz. The d$v(z)$ is the co-moving volume increment from Chapter 3 so that the product d$v(z)$dz gives the volume observed deg^{-2}. To get the number of quasars within that volume, it is multiplied by Ψ, which has units of number per unit volume. It is in the $L(z)$ that complications enter. The first complication is that the observable L for a given survey flux or magnitude limit depends on z through the equations for luminosity distance. The dz bins at smaller redshifts would naturally be expected to show more quasars per unit volume because one can observe deeper into the luminosity function for closer objects. Given a luminosity function $\Psi[L(z = 0)]$, the d$N(z)$ could be determined by calculating the applicable $L(z)$ from equations 3.13 and 3.14 and using this to enter a result for $\Psi[L(z = 0)]$, such as Figure 5.2.

A second complication arises if the possibility of evolution for Ψ is allowed. That would mean that the $\Psi[L(z = 0)]$ is not the same as Ψ for any other z. To take this possibility into account requires a description of how the Ψ changes. It is this description that sets the nature of quasar evolution.

Before proceeding, it is just as well to show in more detail that some kind of evolution is definitely required. Take, for example, the local quasar luminosity function pieced together in Chapter 5, in Figure 5.2. Assume for the moment that this does not evolve and so represents quasars at all z. Using the equation above, $dN(z)$ become as shown in Figure 6.1, taking for illustration a survey limit of 18.3 so as to compare with data from Table 6.1. These results show why evolution of some kind is required. Note that far too few quasars are predicted at higher redshifts compared with the number actually found if there is no evolution. Something has happened to enhance quasars at higher redshift; there are more of them than should be bright enough to detect if scaled from knowledge of local quasars. Our task is to determine the changes in $\Psi[L(z)]$ needed to explain observations extending over all z.

Figure 6.1. Predicted redshift distributions for quasars brighter than 18.3 mag; n has units number of quasars deg^{-2} per redshift interval of 0.2. Histogram: observations of quasar sample from Marshall *et al.* (1983*b*) given in Table 6.1, covering 37 deg^2. Numbers show the actual number of quasars observed in each bin; the smallness of these numbers explains the irregularity of the histogram. Dashed curve: what should be seen for local quasar luminosity function in Chapter 5 without evolution. Dotted curve: what should be seen for local quasar luminosity function evolved as exp $6.3\tau H_o$.

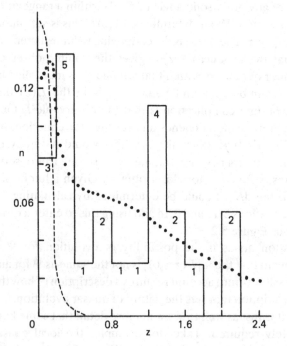

6.3   Luminosity and density evolution

To explain observations such as those just illustrated, the Ψ must be enhanced such that the number of brighter quasars per unit volume increases with z. There is no way to keep up with what is happening to fainter quasars, because they fade below the observational limit regardless of the form of Ψ. If we only need to increase the number of quasars at the bright end of the luminosity function, it could be done two ways. One way is luminosity evolution – often called 'pure' luminosity evolution – in which circumstance the luminosity of all quasars scales up uniformly according to some function of z. In this alternative, it makes no difference where a quasar falls in the luminosity function; its luminosity is increased with z by the same factor as any other quasar. The other alternative of pure density evolution assumes that it is not the luminosity of quasars that is increasing but just their numbers. If the number of quasars per unit volume having a given luminosity increases by a given factor, the chance of observing one increases by that same factor. Under pure density evolution, there are more and more quasars Mpc^{-3} with increasing z; thus, our likelihood of finding the brighter ones is enhanced with higher z, accounting for observed results as in Figure 6.1.

Early analyses of quasar samples were based on pure density evolution (Schmidt 1968, Petrosian 1973, Wills & Lynds 1978). This was an arbitrary but reasonable choice. The quasars in those samples were sufficiently bright and small in number that only the few most luminous quasars could be seen at high redshifts. There are no local counterparts of these very luminous quasars, so there is obviously 'density' evolution for them in that they exist at some z but not at others. How that evolution couples with the remainder of the luminosity function could not be addressed until samples probed fainter into the luminosity function. When dealing with only the bright end of a luminosity function, luminosity and density evolution are difficult to distinguish. It was certainly recognized that pure luminosity evolution could provide an alternative explanation of the data (Mathez 1978).

Other than allowing combinations of luminosity and density evolution, there are no other alternatives to change a luminosity function, because luminosity and density are the only parameters in it. But there are no restrictions on the functional form of either kind of evolution. The point of the observations is to try and restrict these forms. So far, there is still argument as to whether evolution is primarily luminosity evolution or density evolution; improvements in the parameterization of the evolution can be made only with large increases in the number of quasars with known redshifts. Most published models have been argued on the basis of fewer than 200 quasars. Much is yet to be learned. No efforts have been made, for example, to see if the evolution form in any way depends on location in the universe. That is, do the redshift distributions of quasars and so the required evolution depend upon the direction of observation? Perturbations about a mean evolution scenario might someday give important clues to the inhomogeneity of the early universe. As will be emphasized below, virtually nothing is now known about the early phase of quasar evolution, during the first 25% of the history of the universe, even though this is a phase accessible to existing telescopes. Refining our knowledge of quasar evolution will be an effort requiring many years of diligent work at the telescope. When the objective is to understand a process encompassing almost the entire history of the universe, it is a stimulating quest.

For the present, it is adequate to parameterize the evolution with relatively simple forms that should have some physical significance. There are two of these forms, whether applied to luminosity or density evolution. The first is to assume that the evolution, whatever caused it, has been dependent on the passage of cosmic time in an exponential fashion. This is precedented, because so many physical processes undergo exponential decay. Conventionally, the time is taken as look-back time from now, τ in Chapter 3, rather than elapsed time from the Beginning with this form of evolution, $L(z) = L(z = 0) \exp \beta \tau H_o$ for pure luminosity evolution, or $\Psi_z[L(z)] = \Psi_o[L(z = 0)] \exp \gamma \tau H_o$ for pure density evolution. In these relations, β or γ are indices to be fit by observations which determine the amount of evolution. The product τH_o is dimensionless, being the ratio of the look-back time to the Hubble time H_o^{-1}, so only contains terms in z. It is in the relations for τ that one wishes an empty universe with $q_o = 0$ were applicable, because τ is dependent on z in such a complex way for $q_o = 0.1$. The approximation-minded reader might well agree that, since our evolution parameterization is already quite arbitrary, it may prove quite a convenience to adopt the simpler expression for τ in a universe with $q_o = 0$, being $\tau H_o = z/(1 + z)$.

The second alternative evolution form which has often appeared has evolution going as some power of $1 + z$. This is physically meaningful because that relates to the scale size of the universe, so if anything that happens in some way depended on scale size, it might be expected to show up as a function of $1 + z$. Pure luminosity evolution would then have $L(z) = L(z = 0)(1 + z)^\beta$, and pure density evolution has $\Psi_z[L(z)] = \Psi_o[L(z = 0)](1 + z)^\gamma$.

Pure density evolution means that the numbers of all quasars of all luminosities are scaled up by the evolution factor. Pure luminosity evolution requires that quasars with a given density become brighter by the evolution factor. Obviously, combinations of density and luminosity evolution could be tried, as could evolution that did not apply uniformly to quasars of all luminosity. Making the parameterization more complex makes possible wider ranges of fits to the data. Because of the many possible parameterizations for evolution, a correct solution cannot be confidently found until enough redshifts and spectra are available for quasars to determine explicitly the luminosity functions within narrow ranges of z, and then to relate quasars at different z with proper knowledge of the spectrum shape. Sufficient data will require spectra of thousands of quasars rather than the few hundred now available with accurate redshifts and fluxes. Still, a start can be made and, as long as a few assumptions are invoked to simplify the characteristic evolution, important answers can be deduced.

6.4   Comparing local and high redshift luminosity functions

As described in Chapter 5, there are some redshift regimes in which luminosity functions are easier to produce than others. The two favored intervals are the local universe, and the high redshift universe in which quasars are detectable by their Ly α emission. Within the local universe, defined here as $z < 0.1$, many surveys for active galactic nuclei can be combined with surveys for quasi-stellar objects to produce the luminosity function in Figure 5.2, which extends over 12 absolute magnitudes, or a range exceeding 10^4 in intrinsic quasar luminosity. It was emphasized in Chapter 5 that such a combined luminosity function for local quasars can be produced as long as the low luminosity events in galactic nuclei are considered mini-quasars. The many similarities between these events and the more luminous quasars make this an appropriate consideration, satisfying all constraints of the existing data. At the other extreme now available near a single epoch is the luminosity function extending over six magnitudes within the interval $2.0 < z < 2.5$, the most favored interval for detection of faint quasars because of the spectral placement of their emission features. These are not the only luminosity functions that have been deduced for quasars; surveys over all redshifts can be used to synthesize both a luminosity function and an evolution parameterization (Marshall *et al.* 1983*a*, Maccacaro, Gioia & Stocke 1984). The application of this technique will be discussed, but starting with *now* and *then* luminosity functions as described makes for a simple illustration of how evolution can be considered.

What we have in this simple case are two snapshots of quasars in the universe, one at now and the other at an epoch with $z \approx 2.2$, about 10^{10} years ago with the cosmology adopted. After normalizing to the same intrinsic continuum wavelength in the quasar, that of 4400 Å where the blue magnitude is defined for local objects, these snapshots compare as in Figure 6.2. To describe evolution, our task is to determine how the *then* function would have to evolve in order to produce the *now* function. Observational limitations are set primarily by the inability at present to see the faint part of the *then* function. Data shown go to magnitude 21.5, which is the existing limit of reasonably complete surveys. With patient observing it may be possible to push this a magnitude fainter from the ground, but progress beyond that must await HST. For the *now* function, defined within the volume of the universe with $z < 0.1$, the most luminous quasars occur so infrequently that they are not in this volume. That is, once the expected number of quasars Mpc^{-3} drops below the inverse of the volume surveyed, less than one quasar would be in that volume. This limit for local quasars is at $M = -23$. The result of these observational limitations is to constrain the density range within which the *then* and *now* functions can be compared. Other observational uncertainties, such as magnitude calibrations, also enter. Real data from various samples plotted in Figures 5.2 and 5.3

Figure 6.2. Comparison of local quasar luminosity function (*now*) with luminosity function for $2 < z < 2.5$ (*then*), using results from Figure 5.2 and Table 5.4. Outlined zones represent uncertainty in observations. Horizontal arrow: translation from *then* to *now* required for pure luminosity evolution. Vertical arrow: translation from *then* to *now* required for pure density evolution.

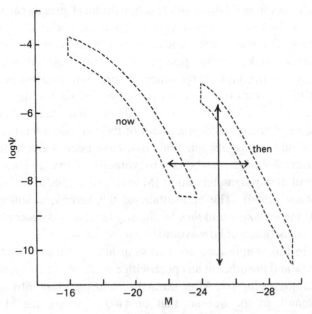

demonstrate the scatter caused by such effects. One of the major challenges for improving the *now* function is to refine the low luminosity end using observations of the broad emission lines to pick out carefully the best luminosity indicator for the 'mini-quasar' surrounded by a bright galaxy. The results in Figure 5.2, while the best depiction of what is available, are only upper limits so will certainly stand major improvement. That can be done in large measure with available facilities.

Taking these results at face value, the futility of allowing complete freedom to the evolution parameterization can be seen. If no constraints are placed in advance on the form of evolution, any part of the *then* function could be transformed to locate it anywhere desired on the *now* function. That would be the allowed procedure for an evolution form that were luminosity dependent, or that combined luminosity and density evolution. It is clear that complex evolution of any form cannot be traced until luminosity functions are available at many epochs between *then* and *now*, so that any part of the luminosity function can be traced through all redshifts. If, on the other hand, we require in advance that evolution be either only in density or only in luminosity, what can happen is much more restricted. Consider first the requirement of pure density evolution. Referring to Figure 6.2, this means that the *now* and *then* functions connect in evolution only by a vertical displacement of about 4.7 in log Ψ. The visible part of the *then* function corresponds to quasars that occur 5×10^4 times more often at $z = 2.2$ than now, so what are now very rare, bright quasars once existed in sufficient abundance to be readily found at $z \approx 2.2$. For the exponential form of pure density evolution, this would imply $\gamma = 15.6$, or a corresponding index of $(1 + z)^{9.2}$ for evolution in $1 + z$. The total number of quasars would have to scale this way.

Neither form of density evolution is physically plausible if we require that quasars always be found, as today, only within the nuclei of bright spiral galaxies ($M < -19$). The local density of such galaxies is $\sim 10^{-2}$ Mpc^{-3} (Felten 1977, Kirshner, Oemler & Schechter 1979); depending on the upper limit, this only exceeds by a factor of $<10^2$ the number of *now* quasars in Figure 6.2. That is, more than 1% of all local galaxies harbor a visible quasar (Section 5.4). If we scale by the required density evolution factor to $z = 2.2$, more than 5 galaxies Mpc^{-3} would be required just to contain the quasars in the universe at that time. That is 500 more galaxies than now exist per unit volume in the local universe, discounting the small irregulars or dwarf ellipticals. The great dilemma with pure density evolution, therefore, is that the faint quasars have to scale by just the same factor as the bright quasars, which requires the existence of more galaxies in the early universe than exist today. The conclusion is inescapable: pure density evolution of the quasar luminosity function cannot be, unless quasars existed 10^{10} years ago within types of galaxies that are no longer visible today.

Consider, therefore, the other extreme, which is pure luminosity evolution. This would require translating the curves in Figure 6.2 horizontally, so that the *then* curve would be transformed to the *now* curve by a fading along the luminosity axis. This could be done with an acceptable fit, and the required amount of fade can be seen to be 4.7 magnitudes, or a factor of 76 in luminosity, applied to quasars of all luminosity. Accepting this gives a simple result, at least for comparing the *then* and *now* nature of quasars. This systematic fading by a factor of 76 could be parameterized using either the exponential or the $1 + z$ form discussed above, with parameters of exp 6.3 $H_0\tau$ or $(1 + z)^{3.7}$.

Such a form of evolution describes the characteristic luminosity decay of the quasar phenomenon. It *cannot* say whether this fading is always restricted to the same set of quasars, or whether different galactic nuclei come and go as quasars whose characteristic luminosities scale with epoch according to the luminosity evolution parameter. How such a decision might be made and other consequences of pure luminosity evolution are discussed in more detail below.

Much improvement would be gained in verifying any deduced form for the evolution if quasars were available at intermediate redshifts. And they are. Most attempts at deducing the nature of evolution have actually come about by considering quasars from surveys covering all redshifts, and then deconvolving the results to produce both a luminosity function and an evolution parameterization. Requiring the determination of a luminosity function in this way introduces still another parameter, so that simplifying assumptions have had to be made. An important one often used is that the luminosity function is represented by a power law; as seen in Figure 5.2, this is a reasonable approximation at the bright end of any known luminosity function for quasars. Undoubtedly, it is not correct at fainter magnitudes. By placing a power law Ψ in the equations above, and hypothesizing pure density or pure luminosity evolution of either exponential or $1 + z$ forms, both the luminosity function and evolution indices could be solved for with a maximum likelihood statistical treatment of the data samples. This was the approach used by Marshall *et al.* (1984) who concluded that pure luminosity evolution of form exp $(7.2\ H_0\tau)$ or $(1 + z)^{3.9}$ were the best fits to the data considered. Note the similarity of these forms to those deduced above just by comparing *now* and *then* functions. We seem, therefore, to have some consensus that a relatively simple form of pure luminosity evolution can account for known quasar samples. Different treatments of the data have produced different conclusions, however, implying both luminosity dependent density evolution for an optical sample (Schmidt & Green 1983) and pure density evolution for a radio sample (Wills & Lynds 1978). Choosing the correct form of the evolution has been a long running discussion, and disagreements naturally arise when data are as sparse as those we have. The last word has not been said, and all of the earlier words should be respected.

The approach to parameterizing evolution which has been described so far utilized for simplicity comparisons of luminosity functions at only two epochs. Most surveys undertaken for quasars yield quasars over a wide range of redshifts. These data can be easily compared with any form of luminosity function evolution through the relation $dN(z) = \Psi[L(z)]dv(z)dz$. To illustrate the fit of my parameterization to such a sample, the data set in Table 6.1 is compared with the exponential luminosity evolution already deduced. The starting point is the local quasar luminosity function from Chapter 5, the *now* function in Figure 6.2. Predictions from the luminosity evolution form found by comparing *now* and *then* functions are compared with this data sample covering other redshifts in Figure 6.1. While the fit is acceptable, it is clear that the small amount of data available precludes overmuch refinement of the evolution parameters.

6.5   Evolution constraints from X-ray properties

While there are already sufficient complications, more arise when comparing the properties of quasars at different wavelength regimes. This could be expected to affect evolution deductions. Quasars may redistribute their energy somewhat with luminosity, the more optically luminous quasars having relatively less X-ray luminosity (Avni & Tananbaum 1982). In such a situation, the evolution of X-ray discovered quasars would not seem quite as steep. Using various X-ray survey samples, luminosity evolution parameters for the X-ray luminosity go as $\exp 4.9\, H_0\tau$ (Maccacaro *et al.* 1984). That is, quasars seem to fade more slowly in X-ray than in optical luminosity. Put another way, whatever characteristic energy generating processes are affected by 'evolution', these processes are such that when quasars become systematically less luminous, they also become relatively stronger X-ray sources compared to their optical fluxes (Kriss & Canizares 1985). Not surprisingly, such conclusions are also based upon applying simplifying assumptions to sparse data. Confirmation must await the direct X-ray observation of quasars at similar redshifts with widely varying luminosities to see if these X-ray to optical ratios hold up. The AXAF would be able to determine the X-ray spectra of all quasars in the *then* optical luminosity function of Figure 6.2. Refinement of X-ray to optical scaling laws and evolution parameterization would then come quickly.

Until more is known about the X-ray spectra of quasars, it remains possible that there is no real dependence of the L_x to L_{opt} on luminosity. Because of the flux limitations of available samples, the quasars with highest luminosity are also those of highest redshift. Most X-ray discovered quasars are very similar in observed flux. As a result, a dependence of L_x/L_{opt} on z would mimic a dependence on luminosity. Such a dependence would come about in the right sense if the X-ray spectrum of quasars were steeper than the optical spectrum. In this case, observing to higher redshift would cause the observed X-ray flux at a fixed wavelength to drop faster than the optical flux at a fixed wavelength (see equation 3.13 and the discussion of the K-correction in Chapter 3). For a quantitative example, consider the conclusion that the optical luminosity evolves as exp 6.3 $H_o\tau$ compared with an apparent evolution of X-ray luminosity with exp 4.9 $H_o\tau$. Both of these conclusions required assuming a power law index for the intrinsic continuous spectra; -0.5 was taken for the optical to obtain the former result.

For a quasar with redshift 2, the evolution parameterizations imply that optical quasars scale up by 67 times in luminosity, whereas X-ray quasars only scale by 26 times. That is, the optical to X-ray luminosity would appear 2.6 times as large at $z = 2$. In order to make this ratio appear to change by this amount, the X-ray spectrum would have to be steeper by 0.9 than the optical. This would produce a factor of $2.6 \approx (1 + z)^{0.9}$. The parameterizers who deduced the form of X-ray evolution assumed an optical index of -0.7 and an X-ray index of -0.5. Had they assumed -0.5 for the optical, this calculation shows that the apparent luminosity scaling for a quasar at $z = 2$ would be mimicked using an X-ray index of -1.2. It is by no means inconceivable that the X-ray spectra of high redshift quasars are as steep as -1.2 (Elvis, Wilkes & Tananbaum 1985). Virtually all X-ray spectra known are for low luminosity Seyfert galaxies rather than high redshift, high luminosity quasars. The lesson is that extensive discussions about the difference in optical versus X-ray evolution of quasars are premature until much more is known about the quasar X-ray spectra. Adopting a flatter optical index and a steeper X-ray index would make the two evolution parameters more similar.

Even if one has no interest in quasar evolution, or does not accept quasars as meaningful probes because of uncertainties in cosmological models, it remains imperative to understand the redshift, flux and spectral distributions of quasars in X-ray bands. This requirement exists in order to remove the quasar contribution to the X-ray diffuse background. Accounting for this background is one of the fundamental problems of X-ray astronomy. There is the possibility that it may be revealing a previously unobserved epoch of energetic events. The X-ray background may have cosmological implications as profound as those which arose from the microwave background. Whether this is so cannot be known until the contribution of faint, known quasars can be removed from the background. Such quasars seem to be the most important component of the X-ray flux from the universe in general (Giaconni *et al.* 1979). Fine tuning of the X-ray evolution and spectral parameters can explain the entire background at particular energies. Because of differences between the observed spectrum of the background and the few spectra available for quasars, it is unreasonable to *use* known quasars to explain all of the background. The background spectrum seems characteristic of a 3×10^8 K thermal spectrum (Marshall *et al.* 1980) while quasar spectra are observed as power laws. Nevertheless, at a given energy, the background flux provides an absolute upper limit to the flux from all quasars at that energy. Some models predict too much X-ray background. Always, the total flux from the background provides a basic boundary constraint to any evolution parameterization, because the integrated quasar X-ray flux cannot exceed this background. Until correct X-ray spectra are known, this constraint is not too helpful in arriving at a detailed evolution parameterization, but it does provide a useful consistency check for the wise modeller.

Typically, this check is made at 2 keV because that is the flux for which most broad band quasar observations are available from the Einstein satellite. This is the X-ray energy to which the optical flux is related via the parameter α_{ox} (Chapter 5), and is thereby the X-ray energy to which optically derived quasar properties are most readily normalized. At this energy, the X-ray background flux is 5.8 keV cm^{-2} s^{-1} sr^{-1} keV^{-1} = 1.2×10^{-29} erg cm^{-2} s^{-1} Hz^{-1} deg^{-2} (Schwartz 1979). This is bright, equivalent to the X-ray flux from a 15 mag quasar. Consider the steps necessary to compare an evolution parameterization with such a limit. We have already encountered the results needed to normalize evolution as described by optical and X-ray quasar samples. It was illustrated above that the X-ray and optical parameterization would be similar if we assumed the X-ray spectrum has a power law index $\alpha = -1.2$. Then, the luminosity evolution parameter is equivalent to the form derived from optical samples: $L(z) = L(z = 0) \exp 6.3 \, H_o \tau$.

The $\Psi[L(z)]$ determined optically for local quasars could be transformed from units of optical absolute magnitude as in Chapter 5 by adopting $\alpha_{ox} = 1.36$. This relates the 2 keV luminosity to the 2500 Å luminosity. To continue to 4400 Å, the blue magnitude at which M is defined, I assumed the optical continuum is a power law of index -0.5. Using the definition of absolute magnitude (Section 3.5.2), the resulting transformation was given in Chapter 5.

The luminosity function in Figure 5.2 for local quasars could be expressed in X-ray units, using the transformations defined. By using this function with the luminosity evolution form applied to an X-ray sample, the $dN(z)$ can be calculated for any X-ray flux limit just as for an optical magnitude limit. When this is done, the resulting integrated flux of quasars at 2 keV as a function of flux limit is as shown in Table 6.2. This table shows that the expected contribution to the X-ray integrated flux from quasars with $z < 2.25$ is less than the observed background. For all $z < 5$, the background is slightly exceeded. Something is wrong, therefore, within the assumptions made so far. It cannot be the cosmological equations, because these simply influence the form of the parameterization. The evolution parameterization can, at its simplest, just be thought of as a convoluted way to *describe* how quasar numbers increase with the flux limit for given cosmological assumptions.

Table 6.2. *2 keV X-ray background from predicted quasar counts*

$f(2\,\text{keV})$ limit units 10^{-32}	quasars deg^{-2} $z < 2.25$	$f(2\,\text{keV})$ deg^{-2} units 10^{-32}	quasars deg^{-2} $z < 5$	$f(2\,\text{keV})$ deg^{-2} units 10^{-32}
373	0.003	1.9	0.003	1.9
149	0.03	9.2	0.03	9.2
60	0.24	31.1	0.24	31.3
24	1.54	85.7	1.59	87.5
9.6	8.8	207	9.3	217
3.8	38	399	42	436
1.53	105	581	142	702
0.61	220	704	370	946
0.24	430	796	796	1130
0.098	800	859	1530	1250
0.039	1310	894	2750	1340
0.016	2070	915	4620	1390
0.0063	3200	927	7270	1420
0.0025	4560	933	10980	1435
0.0010	5050	934	15100	1442
0.0004	5050	934	18100	1444

Total 2 keV quasar background converges to 0.93×10^{-29} erg cm^{-2} s^{-1} Hz^{-1} deg^{-2} for $z < 2.25$, or 1.44×10^{-29} erg cm^{-2} s^{-1} Hz^{-1} deg^{-2} for $z < 5$.

What went wrong? The answer, discussed below, probably lies in seeking the limit for redshift beyond which the evolution parameterization that was deduced no longer applies.

6.6 Utility of radio samples

Although radio-derived samples were the first to be used to demonstrate quasar evolution, there remains a good deal of confusion about what they tell us. The difficulty with radio properties of quasars is that they do not always characterize the same regions of the quasar (Begelman, Blandford & Rees 1984). Compact, 'flat-spectrum' radio components (Chapter 7) generally arise from the immediate vicinity of the nucleus, which is the same volume seen in X-ray or optical radiation. Quasars discovered from these compact sources seem to have comparable evolutionary properties to the other samples (Peacock & Gull 1981), although only a very small fraction of optically detectable quasars are also strong radio sources. More problematic are the 'steep-spectrum' radio sources. These are extended sources, often found tens to hundreds of kpc away from the active nucleus. Visibility of such extended sources depends on the manner in which radio jets, blobs, or plasma from the nucleus react with the surrounding environment. As a result, the apparent evolution of steep-spectrum sources depends both on activity in the nucleus and evolution of the surrounding environment. The overall evolution of steep-spectrum sources may differ from that for quasar nuclei (Wall, Pearson & Longair 1981). Determining the correct parameterization for evolution of quasar radio sources is not handicapped by absence of radio data. The observational difficulty has been in obtaining optical spectra of the associated quasars to determine enough redshifts of faint radio sources. Attempts have begun to relate formally the evolution parameters derived from various sample categories (Danese, De Zotti & Franceschini 1985), but the data remain dismayingly sparse.

6.7 Redshift limit for quasars

No evolution model can be correct if it predicts that quasar densities or luminosities increase indefinitely with redshift. At some epoch in the universe, quasars must have formed. The realistic possibility of finding and tracing this development of quasars is an exciting observational challenge. There are hints that we are already encountering the epoch at which the character of the quasar luminosity function changes dramatically. It is even possible that the 'edge' has been reached, beyond which redshift no quasars will ever be found. These suspicions arise because there are simple evolution parameterizations which are consistent with all available data to redshifts of about 2.2. As already described, a simple form of pure luminosity evolution can connect local quasars to quasars in the redshift interval 2 to 2.5, and also account for those samples covering all redshifts in between. While this is not a unique fit to the data, a major utility of such a fit is to proceed and find just how far in redshift that it works. Of course, such experiments can be done for any parameterization of evolution; a consistent result is that the evolution forms deduced for quasars up to $z \approx 2.2$ do not account for the observations at higher redshifts (Osmer 1982).

In probing the redshift interval 2–5, the observer has the advantage of being able to see the Ly α emission with existing CCD detectors. Because this is the strongest line of all, quasars in this redshift interval should not go unrecognized. Deliberate searches for them have been undertaken (Schneider, Schmidt & Gunn 1983).

Available observations hint that the diminishing of quasar numbers, compared with the expectations of otherwise consistent evolution models, sets in at a redshift only slightly above 2 (Hazard & McMahon 1985). Samples of bright quasars hinted at this initially (Carswell & Smith 1978, Lewis, MacAlpine & Weedman 1979), but it was concluded that the apparent quasar drop off with redshift could probably be accounted for by survey selection effects. Much of the problem in interpreting observations arises because the expected shape of the quasar redshift distribution changes quickly with magnitude limits, even within a restricted redshift region. This is illustrated in Figure 6.3, which shows the anticipated results for evolution vs. observations at redshifts above 2, to magnitude 21. Spectroscopic survey techniques described in Chapter 2 provide efficient ways of finding quasars by searching for Ly α emission. Some such surveys are already available for $2 \lesssim z \lesssim 3.5$ which can provide upper limits to quasar numbers in this interval, just by tabulating all quasars found showing an emission line that could *possibly* be Ly α. Even these upper limits, also in Figure 6.3, fall below the evolutionary expectations. This is the kind of evidence which indicates a change in the nature of quasar evolution in the vicinity of $z \approx 2.2$. Pinning down the form of this change requires many more firm redshifts than are now available from the quasar surveys.

Even if the data show a faster drop off in quasar numbers with redshift than is expected for no evolution, this would not unambiguously mean that the *formation* epoch is really seen. Whether quasars become visible in Ly α as soon as they actually come into existence is another issue. It is possible that quasars are there but have weak emission lines or are shrouded in dust (Hazard *et al.* 1984). The possibility of such effects explains why optical searches alone for quasars need to be supplemented with X-ray, infrared, and radio surveys. Quasar samples defined by these properties would not be subject to the same selection effects operating on the optical surveys so could confirm if the apparent quasar deficiency is showing a literal absence of quasars. Until more sensitive infrared and X-ray telescopes are available, surveys with these techniques will not contribute to this solution. Some estimates of what might be seen are given in Chapter 2. Nevertheless, much remains to be done with the optical surveys that are feasible and whose implementation depends only on the availability of telescope time and enthusiastic observers.

6.8　Physical consequences of quasar evolution

To have a conceptual framework within which to ponder the meaning of quasar evolution, I conclude from the various arguments given that pure luminosity evolution – at least from now back to $z \approx 2.2$ – is as acceptable as anything else, and pleasingly simple. An empirical understand-

Figure 6.3. Shape of redshift distribution for quasars with $z > 2$ and brighter than 21 mag; n has units relative number of quasars per redshift interval of 0.2. Solid histogram: normalized counts of all quasars from Gaston (1983) and Weedman (1985) brighter than 21 mag with emission line that could be Ly α, yielding upper limits to real number of high redshift quasars. Redshift identifications for $2 < z < 2.5$ are more reliable because C IV is often seen, so actual histogram for quasars probably drops off faster than that shown. Curve: Expected shape of redshift distribution if exp $6.3\tau H_0$ luminosity evolution continues.

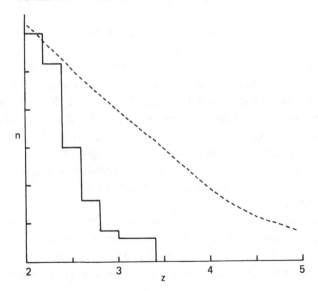

ing of quasar evolution is essential before the explanation of quasar energetics can be complete. The basic energetics equation applying to energy
generation in the most favored model, accretion onto a massive black hole
(MBH) (Rees 1977), is $L = 2.9 \times 10^{46} \dot{m}$ erg s^{-1}. L is the maximum power
available for all forms of energy release for $\dot{m}$ the rate of accretion onto the
hole in solar masses y^{-1}. This follows because mass approaching the hole
approaches velocity c, at which available kinetic energy is $\frac{1}{2}mc^2$. The
maximum $\dot{m}$ which can be achieved depends on the hole mass, as the
Eddington limit sets a maximum to the rate at which a hole can accrete
matter before the pressure of the radiation generated stops the accretion
(Carter 1979). This limit corresponds to luminosity generation of 10^{38} erg s^{-1}
for each solar mass in the accretor. Consequently, we also have $L = 10^{38} eM$
erg s^{-1}, with M the hole mass in solar masses and e the efficiency of energy
generation by accretion relative to the maximum set by the Eddington limit.
For accretion models, e is not likely to exceed a few per cent (McCray 1979).

Regardless of the details of the accretion – spherical, or thick or thin disk
– the bolometric luminosity that can be produced by the release of gravitational potential energy depends only on the three factors: $\dot{m}$, e and M. If
accretion mechanisms are sufficiently similar from quasar to quasar that the
e are similar, M and $\dot{m}$ are the factors that must control evolution. I
emphasize once again that even a complete determination of the form of
evolution does not say how single, individual quasars evolve. What is
determined is the characteristic luminosity of a 'typical' quasar as a function
of epoch, and the conclusions reached regarding evolution say that this fades
by ~ 100 from $z \approx 2$ until now.

There are only two ways to enhance quasar luminosity within these
accretion models. One is to increase M while keeping the radiative efficiency
of accretion constant, so $\dot{m}$ also increases in proportion. The alternative is to
maintain M constant while increasing $\dot{m}$ up to the constraints of the
Eddington limit. Whatever happens, black holes, once formed, must
remain. This means that if high mass black holes existed within galactic
nuclei ten billion years ago, the same or larger holes must still be there. In
the face of declining quasar luminosity over this time span, that leaves little
alternative for models but to assume that evolution is controlled by changing
$\dot{m}$. Finding explanations for such changes is the primary target of physical
models for evolution.

The energetics over quasar lifetimes, based on pure luminosity evolution,
lead to some important conclusions related to the masses of accretors that
would have to exist today in galactic nuclei. There are two alternative
interpretations of pure luminosity evolution. One is that quasar events are

restricted to a small fraction of galaxies which have always contained quasars, the quasar luminosity having faded as the universe aged. The second is that quasars come and go within all large galaxies with the 'on' phase being a few per cent of the galaxy lifetime. Either scenario requires the same total mass accretion, because the total integrated quasar luminosity over the history of the universe is the same. But the former scenario requires all of this mass to accumulate within only a few nuclei, while the latter distributes it over all.

For the most luminous quasars, the required total accreted mass, $\int \dot{m} dt$, is large. From the relations in Chapter 7, a quasar with absolute magnitude -29, about the brightest there is at $z \approx 2$, has $L(z = 2) \approx 10^{48}$ erg s^{-1}. For pure luminosity evolution for this quasar of form $L(z) = L(z = 0) \exp 6.3$ $H_0 \tau$, $L(z) = 1.5 \times 10^{46} \exp 6.3 H_0 \tau$. The integrated luminosity as that quasar decayed from then until now is $\int_0^2 L(z) dz$, which equals 6.4×10^{64} erg. The total mass that must have been accreted to produce this energy release is at least 7×10^{10} solar masses. From the relation between luminosity generation and accretor mass, a mass of 10^{11} solar masses had to be present in the accretor at $z = 2$, for e of 10%. The sum of that plus the mass accreted since $z = 2$ means that this luminous quasar would by now contain a black hole of almost 2×10^{11} solar masses! If the luminosity has decreased by a factor of about 100, while the accretor mass has increased by a factor of two, luminosity evolution can only be accounted for if e has by now dropped to 0.05%, or 1/200 of its original value.

For the least luminous quasars in Figure 6.2, those with present-day $M = -16$, these numerical values scale down by a factor of about 10^4. Using the same reasoning as before, this means that the lowest luminosity quasars in the local universe would reside in galactic nuclei containing accretors of at least $10^7 M_\odot$.

There is a way to reduce the required accreted masses by a factor ~ 100 for all quasars. This is to assume that all large galaxies at some time in their lives contain quasars. As discussed in Chapter 5, the co-moving space density of local quasars is a few per cent of that of local bright galaxies. If, therefore, quasars come and go in all such galaxies, the duration of the quasar events must total only a few per cent of the entire galaxy lifetime. That would reduce the total accreted mass by a factor between 10 and 100 compared with having quasars always restricted to a small fraction of galaxies. In this scenario, the total accreted mass never compares with the mass M of the primordial accretor, although the initial mass of the accretor required to trigger the brightest quasar event ever occurring is not reduced within this alternative.

Combining all the preceding reasoning leads to the requirement of substantial accretor masses in galactic nuclei. Depending on whether a galaxy has harbored a bright or faint quasar and whether quasar events have been distributed among all galaxies, this accretor mass in galactic nuclei today could be as small as $10^6\ M_\odot$ or as large as $10^{11}\ M_\odot$.

Today's quasars must be powered by black holes at least as massive as those in the past, so the accretion efficiency today must be much less. Models of the accreting region have as a critical objective the task of determining the mass and efficiency of the accretor (Chapter 9). The scenario described leaves the formation of the initial hole as an essentially primordial process, one that predated the first quasars observed. Whether or not the quasar phenomenon has affected every galaxy or is a perturbation on only a few can be determined by learning if there are dormant holes of the necessary size in all nuclei. Is the existence of a quasar dependent simply on whether there is sufficient material to feed the hole? Are the holes lying in wait in the nucleus of every galaxy? If so, it means that the form of evolution we have seen represents the statistical flaring of short lived quasar events in many nuclei rather than the slow decline of such events in a few nuclei that have been quasars from the start. We may learn the answer with observations from HST, which can measure the gravitational potential in the nuclei of non-active galaxies.

6.9   Searching for quiescent accretors

In principle, the mass of a gravitationally bound system of particles can be determined by observing their kinetic energies. Assuming the system is in equilibrium and has no net rotation, the virial theorem equates $2\mathrm{KE} = M\langle v^2\rangle = -\mathrm{PE}$. The system is supported against collapse by the orbits of its individual members about the center of mass. v represents the average velocity of such members and M their total mass. For a system of stars in the nucleus of a galaxy, the virial theorem can be assumed to apply. The average velocities are observable. What is really observed is the radial component of velocities: the line-of-sight velocity dispersion σ. The σ is the one-dimensional projection of three-dimensional motion. Because of this, $v^2 = 3\sigma^2$. Because of the varying radial velocities of the many individual stars in the system, the observer sees a spectral line which has been artificially broadened by the combined effects of these motions. The line profile is gaussian in shape; in velocity units $I(\Delta v) = I_0\exp(-\Delta v^2/2\sigma^2)$. A commonly measured parameter for any spectral line is the full width of the line at half maximum intensity (FWHM), so FWHM $= 2.35\sigma$.

The potential energy of the system is $PE = -G \int_0^R M(r)r^{-1}dM$, $M(r)$ being the mass inside radius r and dM the mass in a shell at radius r. For density ρ, $M(r) = 4\pi\rho r^3/3$ and $dM = 4\pi\rho r^2 dr$. If the particle distribution was homogeneous (not the case for real self-gravitating systems), $\rho \neq \rho(r)$ and $PE = -3GM^2/5R$. Equating via the virial theorem would yield $3GM^2/5R = 3M\sigma^2$; the mass of the system could be solved for by observing R and σ.

This approximation is not applicable to real systems, but it does show the proportionalities among the variables and demonstrates how, in principle, the mass of a system can be determined. If one thought that the system were made of stars of known luminosity, the mass of stars present could be estimated by the luminosity of the system. If that stellar mass were less than the measured system mass, there would be evidence for another component, such as a massive but dark accretor. This is the fundamental concept which can be used to probe for dark mass within real stellar systems.

This measurement is done by observing the stellar spatial and velocity distribution in nuclei. It requires measuring the stellar component only, so confusing light from an existing quasar is undesirable. In fact, it will not be possible to use this technique on many active nuclei for this reason. The most important active targets are the nuclei with sufficiently low level activity that the stars can still be seen, but which still must have mini-quasars. The priority target is M81 (Chapter 9). Galaxies with no evidence of nuclear activity, such as M31, will show if there is or is not a dormant quasar in each. If no evidences of such are found, it will imply that activity has always been where activity is still seen, so we could conclude that only a minority of galaxies ever developed the nuclear accretor.

The observational techniques for searching for central mass concentrations in galactic nuclei are summarized by Young *et al.* (1978) and Sargent *et al.* (1978). The spatial distribution of the stars can be modelled to see if there is an unexpected cusp in the center, arising from a higher concentration of stars than expected. The velocity dispersion in this cusp is measured to determine if it confirms the presence of more mass than can be accounted for by the stars. Observations are extremely challenging even for quiescent nuclei because of the high spatial resolution required. A relation for comparing parameters is $\sigma^2 = 4\pi G\rho_0(r_c/3)^2$, an accurate representation of the virial theorem for real systems. The ρ_0 is the central stellar mass density, of order 100 $M_\odot$ pc^{-3}, and r_c is the core radius (radius at which the surface brightness of starlight drops to half its central value), of order 100 pc. Searching for stellar luminosity cusps or increased velocity dispersions within the nuclei of nearby galaxies is a priority objective for HST.

In Chapter 9 on nuclear structure, what can now be said about the mass of the accretor is reviewed. All that are available are upper limits, which do not change obviously with the luminosity of the quasar, and which are adequately high to account for differences of 100 in luminosity just by accretion efficiency. All of the arguments available now are at the level of basic consistency arguments. They cannot be said to prove anything. As discussed in Chapter 9, the very fundamentals of this entire hypothesis – whether the luminosity is really due to a massive accreting black hole – are still unsatisfyingly vague. To do the best with what is available, and to construct an evolutionary scenario that is at least consistent, it makes sense to attribute the cause of evolution to consequences expected to arise in a galactic nucleus. Firstly, the hole must be fed. It is not easy to get gaseous matter into the vicinity, as shown by the evidence for systematic outflow of the gas that can be seen in the nucleus. But stars are always in galactic nuclei, some in orbits that carry them right through the center. So an attractive source of matter to feed the hole is some of these stars, broken up by tidal destruction caused by the hole itself, or by collisions with other stars. If such are presumed to provide all of the raw materials for accretion, elaborate evolutionary scenarios can be constructed that have quasar luminosity essentially controlled by the density of the stellar distribution in the nucleus (Cavaliere *et al*. 1983).

An independent and intriguing alternative to account for evolution by varying the feeding has recently surfaced from a wide variety of observations. These data show a tendency for both nearby quasars (within the nuclei of visible galaxies) and distant quasars (for which the underlying galaxy is not conspicuous) to be found in interacting systems (Hutchings, Crampton & Campbell 1984, Dahari 1985). This is another area in which the patient gathering of observational statistics is a pressing objective, underway and doable, with substantial ramifications (Keel *et al*. 1985). The interaction process is known to disturb galaxies tidally, affecting the orbits of both gas clouds and stars. Is it this disturbance that removes enough angular momentum for material to settle to the nucleus? If interactions control the feeding of the nucleus, evolution would then depend on the rate of interactions. Why should this have been greater in the past (Roos 1985)?

If future observations show that all galaxies contain MBH, this result would not conclude the quasar story. Because the presence of that MBH must be explained. Did the hole come as a consequence of the stars, as can be extrapolated from core collapse models? Or was the hole there before, and was it the seed around which galaxies formed? Some have been discouraged by implications that quasars may show us nothing more than the extremes of gravity in our universe. But we are far from understanding how gravity shaped the distribution of matter as we know it, how the galaxies and clusters of galaxies collected. Even if no new forces and no new physics are required for quasars, nothing more than the gravity of the hole, there is still much to ponder. Even more exciting from the standpoint of exploring a scientific mystery would be indications that there are *not* massive condensed objects in galactic nuclei. Without them, theories to explain quasars would be back to square one.

7

CONTINUOUS SPECTRA

7.1 Introduction

Quasars are unique among objects of the universe in the observable span of their continuous spectra. In some cases, the same quasar can be seen with existing instruments at wavelengths from X-rays to radio, including everything in between. The quasar continuous spectrum is deceptive. Order-of-magnitude agreement over all wavelengths, from tens of centimeters to fractions of an angstrom, covering a range of $>10^{11}$ in frequency, can be obtained by fitting a single power law spectrum, of form $f_\nu \propto \nu^\alpha$, where α is ~ -1. It is tempting in the face of such a result to attribute all parts of the spectrum to related mechanisms. As has become very clear from more careful examination of spectra, that is not valid. Different components of the continuous spectra are produced by drastically different mechanisms, and there are sometimes no physical relations among these mechanisms. It is nevertheless assumed that all of these mechanisms are basically set in motion by a single underlying engine, such as gravitational accretion, but the radiation which comes out represents many ways of transforming gravitational to radiative energy. The greatest success of the intensive observational effort has been to show the exceptional similarities among spectroscopic properties for quasars covering a factor approaching 10^7 in luminosity. This is the single key fact to be explained by theoretical models of quasars. Whatever processes control the radiation must be capable of scaling over this range of energy release without fundamentally changing character.

Attempting to understand all of the radiation mechanisms that could conceivably operate in the complex environment of a quasar has led to comprehensive discussions of such mechanisms (Tucker 1975, Rybicki & Lightman 1979). Many of these mechanisms will be reviewed briefly in this chapter. Primary effort will be given to describing the observational parameters relating results at various frequencies. An immediate source of confusion is notable. Observers at different wavelengths have different conventions for describing those wavelengths: X-ray astronomers use energy; those working in the ultraviolet and optical use wavelength in angstroms; infrared astronomers use wavelengths in micrometers; and radio astronomers use frequency. It is too late to enforce a change in these conventions, so this mixture of units will be accepted. I will refer to them interchangeably, without further apology, as that reflects the circumstance in the available literature. For reference and convenience, Table 1.1 brings together the various conversions among units and the definitions of fluxes measured, all tied to the flux measurement of erg $cm^{-2} s^{-1} Hz^{-1}$, which is used in the cosmological equations of Chapter 3.

Discussing the various radiation-generating mechanisms could be approached in order of wavelength, or in context of the various physical mechanisms responsible. There is no simple correlation between the two. All radiation mechanisms involved are either thermal or non-thermal, there being no other alternative. But each of these categories contains several different processes taking place in quasars, intermingled among wavelength bands.

7.2   Thermal processes

Thermal radiation arises from energy levels excited by collisional processes among particles that have achieved a statistical distribution of velocities. These velocities are non-relativistic, of order 10^6 cm s^{-1} if $T \approx 10^4$ K. The energy distribution of the particles is a distribution characterized by a single temperature; that is, once the temperature is stated, it is defined how the kinetic energies of the individual particles in the entire collection are distributed. Radiation arising from collisional interactions among these particles within an object sufficiently dense to be optically thick has an emergent radiation spectrum characterized by the Planck function. Such black-body spectra from hot, optically thick gases or solids can be found in several regions of the quasar continuum. If a sufficiently large wavelength range is observed, it may be possible to demonstrate the spectrum shape of a Planck curve at a single T. The only other unambiguous demonstration that thermal mechanisms of some kind are present is to observe absorption or emission features in a spectrum. These arise from equilibrium processes which can be characterized by an excitation temperature.

Hot, optically thick accretion disks or dense clouds are suggested as sources of a black-body continuum which is observed to be strong in the ultraviolet, often called the 'blue bump'. Stars, of course, are the most widespread thermal black bodies. In some cases, particularly the fainter quasars, stellar radiation from the volume surrounding the nucleus contributes to the continuum observed. The final significant black-body contributions are in the infrared, but not from optically thick gases. Infrared radiation in quasars can come from solid particles ('dust'), which have been heated by absorbing higher frequency radiation and which re-radiate this absorbed energy in a continuum defined by the temperature of the particles. If only these various black-body radiation mechanisms were present, quasar spectra still could be very complex because various components could have different temperatures. Just the possibilities from the dust alone are endless, describing various dust geometries containing zones with different temperatures. The resultant spectrum would not be that of a Planck curve at a single temperature, which is the characteristic by which the black-body spectra of stars have been conventionally recognized.

The other common source of thermal continuum radiation is from an optically thin gas and is the free–free continuum, or thermal bremsstrahlung. This arises from interactions between particles that do not produce recombinations. Free–free radiation may dominate the spectrum at some wavelengths, being particularly conspicuous in the X-ray for very hot gases ($T \gtrsim 10^8$ K) and in the radio for gases of $T \sim 10^4$ K. Free–free radiation has an emission coefficient dependent on exp $-h\nu/kT$. Except in the X-ray, a wide enough spectral range is not observed to specify thermal bremsstrahlung just on the basis of spectral shape. Even in X-rays no quasar spectra are known precisely enough to specify thermal bremsstrahlung unambiguously just from the shape. Source variability, discussed below, is a strong argument against a thermal X-ray source. If the gas becomes optically thick, photons generated by free–free interactions cannot escape without reprocessing, and the emergent spectrum will take on that of a Planck spectrum. When a spectrum is quoted as free–free, the implication is that it is optically thin.

Another thermal continuum that can occur at restricted wavelengths is that from recombinations, particularly the Balmer continuum. This is closely correlated with emission line properties. In addition, there are pseudo-continua that can arise from blends of many closely spaced emission lines, particularly Fe II lines, that confuse the ultraviolet and optical spectra, but which do not arise from true continuum processes.

7.3   Non-thermal continua

Non-thermal radiation arises from particles, usually moving at relativistic velocities, which have not achieved an energy distribution as a consequence of mutual collisions. There is not in this case a single 'temperature' to define the energy distribution of the particles. The energy distributions are reflective of the initial acceleration mechanism, and not of subsequent collisional processes by which the particles share their energies. Many forms of energy distribution are thereby possible, leading to a correspondingly broad range of energies for any radiation produced by these particles. It is common, however, to characterize the energy distributions of particles in a non-thermal ensemble by a power law, such that $dN(E)$, the number of particles at a given energy, is $dN(E) \propto E^{-\gamma}$.

Because it has been pondered the longest, the best known form of non-thermal radiation is synchrotron radiation. This is produced by the loss of energy from electrons moving relativistically through a magnetic field. (Synchrotron radiation can also arise from protons, but this mechanism is considered so infrequently that it is referred to as proton synchrotron radiation.) For single values of the magnetic field it is reasonably easy to show that electrons with such an energy distribution would produce radiation whose emission coefficient is a power law, $j_\nu \propto \nu^{0.5(1-\gamma)}$. Synchrotron radiation was the first non-thermal source of radiation definitely observed, accounting for the radio flux of many extended sources, including the lobes of radio galaxies, the disks of spiral galaxies, and supernova remnants. All such cases are sufficiently extended to be optically thin, and the radio flux from optically thin, non-thermal sources is quite uniform in spectral shape. F_ν is proportional to ν^α, with $-0.5 < \alpha < -1.0$ in most circumstances. This uniformity implies that whatever acceleration process is responsible for the relativistic electrons, injects those electrons with similar energy spectra in many different environments. Optically thin synchrotron radiation is, therefore, recognizable by an output radiation spectrum that is a power law, and any such spectrum is usually attributed to this mechanism. In large measure, because of this simple result for synchrotron emission, spectra observed to be of power law shape are often assumed as non-thermal. This is not necessarily correct because a suitable mixture of thermal spectra at different temperatures can produce a resultant that is a power law. There can be no doubt, however, that some parts of quasar spectra are non-thermal (Jones, O'Dell & Stein 1974). This is a conclusion of basic importance that must not be obscured by the complexities of interpreting non-thermal spectra. What it means is that quasars contain accelerators that can produce relativistic particles. Underlying all of the attempts at understanding the radiation mechanisms is the effort to understand how these accelerators work.

Even if synchrotron radiation from electrons with a uniform energy spectrum is the only source of radiation at all wavelengths, the entire spectrum need not have a single power law. The reason is that synchrotron emitting volumes have different optical depths at different frequencies, and self-absorption changes the output spectrum. For example, the electrons in a cloud of plasma that are capable of radiating photons can also absorb or scatter them by the Compton process. This means that a generated photon may not escape before interacting with another relativistic electron. Often the photon energy is boosted by this subsequent interaction – the synchrotron self-Compton (SSC) mechanism. This is a useful way for producing high energy photons from initially lower energy photons, and can even make X-rays out of photons at radio frequencies (Condon *et al*. 1981). Nevertheless, the initial photon disappears so the plasma is optically thick to it. In circumstances where photons are generated in dense environments, the output depends on the path length through a cloud for a given photon, so the observer finally sees radiation from an ensemble of photons that have come from regions of varying optical depths. Such optically thick synchrotron radiation can have an apparent power law much flatter (larger α) than the actual spectrum which the radiating electrons generate. The reason is just that the lower frequency photons are absorbed more easily. Often, compact radio sources show $0 < \alpha < -0.5$. For the same reason, it is easier to get optically thin synchrotron radiation for higher energy photons, which is why optical, ultraviolet or X-ray synchrotron radiation from a compact source may show a single steep power law when the radio spectrum does not.

Because radio photons that are absorbed out are often boosted to higher energies by this Compton scattering off the electrons, they can reappear in the X-ray, ultraviolet or optical with a distribution showing the same power law as the original photons. This is one reason why it is difficult to distinguish between pure synchrotron radiation and synchrotron self-Compton radiation. This confusion is most serious in the X-ray continuum. On the other hand, the observation of an X-ray spectrum with a slope of -0.5 to -1 is taken as empirical evidence that such radiation is arising non-thermally, from an ensemble of electrons with the same energy spectrum as seen in optically thin synchrotron sources at radio or infrared frequencies. The fundamental energy source – the relativistic electrons – is the same whether the X-ray photons seen are primary or secondary.

Just because an X-ray, ultraviolet, optical or infrared spectrum is a power law is no guarantee that it is non-thermal. Power law spectra can be produced in all of these regimes by suitable, if *ad hoc*, combinations of thermal sources. This is not a source of confusion unless there is reason to believe widely varying temperatures might be encountered in the same source, but models of accretion disks and distributions of dust have just this property. So the dispute is not settled within all wavelength regimes as to how the observed power law spectra of quasars arise, from a fundamentally thermal or non-thermal source.

7.4   Relativistic beaming

Once it is realized that matter can move at relativistic velocities within quasars, it is necessary to face the possibility of relativistic corrections to what is observed (Blandford & Rees 1978). One of the most basic reasons for observing the continuous spectrum is to determine the luminosity being radiated by the source, so the energetics can be examined. Doing this requires the assumption that flux is not directionally dependent, so that the equations of Chapter 3 relating flux and luminosity can be applied. An extraordinary thing about quasars is that this assumption is not always correct. Radiation is not necessarily isotropic. Unquestionably, relativistic particle velocities are present in some radiating quasar plasmas. As long as the particles have a random distribution of direction for their motion, radiation produced thermally or non-thermally leaves isotropically when integrated over the whole radiation source. Subsequent absorption processes may arise anisotropically, as the photons attempt to escape, but that anisotropy is not the issue. The most notable anisotropic effects arise from radiating particles in a cloud moving coherently (bulk motion) with a relativistic velocity. In this case, any radiation produced from the cloud is concentrated in a cone projected in the direction of the motion. An observer in that direction would see an enhanced luminosity, and the observer's perception of time scales in the cloud would be altered. Meaningfully enough, this process is called relativistic beaming.

The basic equation for relativistic beaming describes how the source luminosity is enhanced in the direction of motion. Terms must also include a description of the radiated spectrum, because the observer sees the continuum blueshifted relative to the source reference frame. In effect, these are the same terms but in the opposite sense that showed up when calculating cosmological equations for redshifted sources. The equation which has been used (Scheur & Readhead 1979, Urry & Shafer 1984) describes the factor relating observed luminosity of the beamed source to the luminosity that would have been observed in the absence of source motion. Calling this the enhancement factor E_ν, $E_\nu = [\gamma(1 - \beta \cos \theta)]^{\alpha-3}$. The parameter γ is the customary Lorentz factor $(1 - \beta^2)^{-0.5}$, and β is the ratio of source velocity to the speed of light. α is, as usual, the index of the power law spectrum, and θ is the angle between the line-of-sight and the velocity vector of the moving source.

For a few cases, the extraordinary resolution capabilities of VLBI have made it possible to observe the bulk motion of relativistically beamed sources (Cohen *et al.* 1983). These are the so-called 'superluminal' sources. The γ factors in such cases are of order 5, giving $\beta = 0.98$, so E_ν can be very large. For these values with $\theta = 5°$ and $\alpha = -0.5$, $E_\nu = 1.7 \times 10^3$. What this means is that a single relativistic cloud that happens to be moving toward the observer can easily dominate the integrated luminosity from the rest of the quasar. Furthermore, the time scale of changes in the source would be diminished by a factor $\gamma(1 - \beta \cos \theta)$, making variability appear much more rapid than was intrinsically the case. For this example, the time scale changes by a factor of 8.4. Only small deviations in θ would be required to change the apparent luminosity dramatically. If θ changed from 5° to 6° in this case, E_ν changes to 1.4×10^3; the source would suddenly diminish in luminosity by about 20%.

Non-thermal continuous radiation in any part of the spectrum can be enhanced by relativistic beaming (Henriksen, Marshall & Mushotzky 1984). This is very troublesome, because it means that the luminosity or variability characteristics observed in a given quasar may not be representative of anything other than the particular cloud moving toward us. It would be analogous to trying to understand the basic structure of the Sun by observing a single solar flare. Only for thermal radiation can we be confident that the luminosity is isotropic. This is why emission lines are the most reliable luminosity indicators. Using the emission lines, an indirect argument can be made that the non-thermal continuum from the great majority of quasars *is* isotropic and so not dominated by one or a few beamed components. This argument is that the relative strengths of emission lines compared with those of the continuum are very similar in most quasars. If continuum variability is observed, there is never a guarantee that this is isotropic unless it is eventually followed by comparable variability in the emission lines (Ulrich *et al.* 1984).

Nevertheless, the occasional effects of beaming are so strong that an entire class of quasars exists based upon observational properties seemingly a consequence of beaming. These are the 'blazars' (Angel & Stockman 1980). Among these are 'optically violent variables'. These are quasars with emission lines whose luminosity remains constant while the continuum varies dramatically, by factors as much as 100. The most extreme example of blazars are the BL Lacertae objects, with no detectable emission lines (by definition of this class of objects). They are characterized by rapid variability, simple power law spectra over a wide wavelength regime extending into the radio and high polarization (Cruz-Gonzalez & Huchra 1984). The latter two properties would be characteristic of synchrotron radiation from single sources, the first can be explained by the simple hypothesis that a synchrotron emitting plasma is approaching the observer at relativistic velocity but slightly varying angle.

Blazars and BL Lacertae objects represent only about 1% of quasars (Impey & Brand 1982, Maccacaro *et al.* 1984). The remaining quasars show strong emission lines that are reasonably constant compared with the continuum. This leaves the puzzle of how the few beamed sources differ. Would all quasars appear this way to an observer in some direction? Are blazars the small fraction that happen to have relativistic clouds moving in our direction? The extreme alternative is that blazars have relativistic bulk motions in many directions, so any observer would see them as we see them, but other quasars have no such motions visible from any direction. In support of the latter suggestion is the morphological difference between quasars and BL Lacertae objects; the former are in spiral galaxies but the latter in ellipticals (Miller 1981). Gas in the quasar nucleus can produce emission lines and reprocess synchrotron radiation, so a gaseous nucleus might prevent beamed synchrotron radiation from escaping in any direction.

Unfortunately, any non-thermal radiation mechanism, because it is characterized by relativistic velocities, is subject to suspicion concerning its isotropy. The possibility of beaming makes arguments about source sizes and luminosities weak. Source size deductions are fundamental to hypotheses about quasar structure (see Chapter 9). It is preferable if these deductions are based upon emission line properties, so there are no beaming suspicions. A troublesome situation arises in desiring to use variability time scales to set the size of the continuum-emitting region. If this continuum is non-thermal, beaming is a possibility and, if present, implies smaller size scales than really exist for the bulk of the luminosity generated isotropically. It is reasonable to assume that a source which shows no significant variability, or at least that part of the flux that is not variable, represents an unbeamed component. The lesson is never to trust an unstable non-thermal source for determining quantitative, steady state parameters.

7.5 Observed generalities of quasar continua

To a sufficient approximation that it bears discussion, the continuous spectrum of any quasar can be considered as $f_\nu \propto \nu^{-1}$ over all frequencies at which this spectrum is seen. No quasar really has precisely this shape, and no quasar really has a single power law spectrum over all observable frequencies. But this approximation is nice to start with, and to compare reality with, because a spectrum of this shape has a simple property regarding energetics. Integrating f_ν between any frequency limits shows that the total flux, and so the total luminosity, is the same within each decade of frequency range for $f_\nu \propto \nu^{-1}$. That is, the flux between 1 and 10 Å would be the same as that between 1000 and 10 000 Å or between 2 and 20 cm, etc. That quasars can come close to this approximation then gives an easy proof that they generate comparable amounts of luminosity in the various regions of the spectrum described as X-ray, optical or infrared. The major difference among quasars is whether they also show a continuation of this power law spectrum into the radio. 'Radio strong' quasars, particularly the blazars, show a continuum extending to meter wavelengths. But 'radio quiet' quasars die in luminosity somewhere shortward of the millimeter band. The large majority of known quasars are so radio quiet that they are not detectable with the most sensitive radio telescopes. As a result, the radio properties of quasars cannot be discussed uniformly; trying to understand why they differ so much is one of the major puzzles, pursued more in Section 7.10.

Note that a spectrum 'flatter' than ν^{-1}, i.e. with $\alpha > -1$, carries relatively more energy at shorter wavelengths, with the converse for a spectrum 'steeper' than ν^{-1}. Given the convenience of using power law continua, spectral shapes are often described by an artificial α. For example, α_{ox} is defined as the power law slope that would be required to connect X-ray observations at 2 keV with ultraviolet observations at 2500 Å (Tananbaum *et al.* 1979). If the real continuum had $f_\nu \propto \nu^{-1}$, obviously α_{ox} is -1. (Readers should be aware that there is no firm sign convention for the power law index in a continuous spectrum. It is often the case that α is defined by $f_\nu \propto \nu^{-\alpha}$. It is always necessary to confirm the convention in each case. Authors are reminded to make this clear.)

7.6   Infrared continuum

Looking at the spectra of quasars to consider the departures from a single power law, we can hope to isolate those parts of the spectrum in which related physical mechanisms apply for the luminosity generation. Enough has been said for the moment about the absence of consistency in the radio, so the discussion proceeds starting with the infrared, where most quasars do seem quite similar. Infrared spectra have been routinely observed as longward as 100 μm (IRAS). From the longward limit down to 2 or 3 μm, infrared spectra can be fit by single power laws. A simple lesson in estimating the power law index is given by comparing the optical B magnitude with the IRAS 60 μm fluxes for the Seyfert 1 galaxies in Table 5.1. Using the unit transformations in equation 3.15, $\alpha(B - 60\,\mu m) = -9.1 - 0.19\,B - 0.47\log f(60)$, for $f(60)$ in erg cm^{-2} s^{-1} Hz^{-1}. The mean value is 1.11, which is consistent with the index of nearly -1 found for other infrared wavelengths

(Miley, Neugebauer & Soifer 1985, Soifer *et al.* 1983). Of course, this fitting process is greatly simplified when there are only two or three data points in the entire spectrum. It is not proof that a single energy generation mechanism applies; bumps and wiggles in the spectra are seen when objects are sufficiently bright for high spectral resolution. Nevertheless, ν^{-1} comes pleasantly close to a descriptive fit for the infrared data. Steeper spectra can have two alternate, but extremely different, explanations. One is cool dust, re-radiating energy absorbed from shorter wavelengths. Depending on the distribution of dust temperature, spectra can have varying shapes. In a few cases, spectral absorption features attributable to dust grains have been observed. An alternative explanation for steep infrared spectra applies to a rare class of very red blazars, which appear to have optically thin synchrotron radiation from electrons with a steep power law energy index. The steepest infrared spectrum observed from such an object approaches ν^{-3} (Rieke, Lebofsky & Wisniewski 1982).

Even though most infrared quasar spectra have little scatter about the ν^{-1}, these examples illustrate that this similarity has not led to reconciliation concerning the source of infrared luminosity. Cogent arguments are presented favoring synchrotron radiation and, even for the same objects, favoring re-radiation from dust. Only in the case of the blazars, with such rapidly varying fluxes as to require beamed synchrotron radiation, can the issue be considered settled. One's preference in interpretation makes no difference when using the infrared results to determine the total luminosities of quasars, but the issue of whether there is or is not hot dust in quasars affects modelling of the emitting regions (Chapters 8 and 9). As a general rule, the smoothness of the infrared spectra along with the variability and polarization characteristics leave the impression that it is primarily non-thermal (Stein & Soifer 1983). Nevertheless, a few dusty quasars reveal themselves via silicate spectral features from the dust (Cutri, Rieke & Lebofsky 1984), so there is no doubt that dust can be associated with quasars (Rudy 1984).

The problem of disentangling thermal and non-thermal sources becomes even more difficult in the near infrared (1–3 μm) and the optical. Now, we not only have to worry about hot dust as a thermal source but also stars of varying temperatures. Planck curves show readily that stars with temperatures of a few thousand kelvin, of which there are many, will certainly be important radiation sources at these wavelengths. Hotter stars become progressively more important into the optical and ultraviolet. At least for stars, characteristic spectral absorption features are known, and highly sophisticated deconvolution programs have been developed to correct for stellar continuous spectra on the basis of observed spectral features (Shuder 1981). But dust cannot be taken out in this way; the non-thermalists argue that dust is of little importance below ~3 μm, just because characteristic Planck temperatures then exceed 1000 K, above which the survival of dust grains is questionable.

7.7   Optical and ultraviolet continua

Many continuum generating mechanisms are feasible in the optical and ultraviolet, so it is no surprise that quasar spectra are not all the same at these wavelengths. Neglecting the complications from varying mixtures of stars of different temperatures, we have as possibilities: optical synchrotron radiation; free–free and bound–free continua from ionized hydrogen; blends of iron emission lines that mimic a continuum; and thermal black-body radiation from optically thick sources that are not stars, such as accretion disks. Furthermore, any of these mechanisms can be affected by absorption by intervening dust so that the emergent continuum does not have the same shape as the intrinsic continuum. Various self-consistency conditions can be applied to deduce the true spectra in the optical–ultraviolet regime. For example, there must be sufficient ionizing photons shortward of 912 Å, the Lyman limit of hydrogen, to account for all of the ionization necessary to produce observed hydrogen emission lines. Matching continuum and emission lines in this way can yield constraints on the structure of the ionized volumes. Ionization models that explain the emission lines must be consistent with the ionizing portion of the continuum. All in all, the ionization models are consistent with the existence of a power law continuum through the ultraviolet with slope -0.5 to -1.0 (Davidson & Netzer 1979).

At blue and ultraviolet wavelengths, quasar spectra in general are flatter than anywhere else in the spectrum. The $f_\nu \propto \nu^{-1}$ power law begins to flatten at about 5000 Å, and various broad features show in the continuum that are not simple power laws. From 5000 to 1000 Å, it is reasonable to approximate the spectrum with $\nu^{-0.5}$ (Richstone & Schmidt 1980). Unfortunately, knowledge of the spectrum shape in this regime is particularly critical. It is in this wavelength range that quasars are seen at redshifts up to ~3 with most broad band optical surveys. To determine quasar luminosity functions and parameterize evolution, it is necessary to infer the monochromatic luminosity at rest wavelengths other than that actually observed. Simple relations make this possible (Chapter 3) for power law spectra; such calculations have not been made using more realistic spectra. For this reason alone, better empirical knowledge of the optical–ultraviolet quasar continuum is important.

The simpler the model, the more attractive it is. It is desirable to explain as much of the continuum as possible with a single mechanism so the physics of the luminosity source can be isolated. It is always possible to reproduce any observations by incorporating many parameters, all of which are adjustable until the data are fit. But that is not clever. Such models are not unique. Someone else may weight the parameters slightly differently and produce an equally satisfactory fit. The fewer mechanisms that have to be mixed to produce self-consistent continua, the more credit is due to the modeller. This procedure is not recommended just because it is simpler and makes the results easier for everyone else to understand. More fundamentally, we are after the basics. Probably, a little bit of many things is going on in quasars. What is important for our current efforts is to learn those mechanisms that are most significant, and concentrate on understanding them. We are simply not far enough along to account for all of the details.

A useful piece of work in the spirit of this approach has been the effort to deconvolve the infrared–optical–ultraviolet continuum into a two-component model. This had to be the next step after it was clear that a single power law did not fit this entire spectral range. Trying to fit with the latter was a good attempt because it could be done with a single-source synchrotron model. That still works with the data for blazars, but quasar spectra show a definite departure from a single power law in the optical and ultraviolet. This departure is in the sense that the power law from the infrared begins to flatten in the optical and does not plummet down steeply again until the far ultraviolet. This property of quasar continuous spectra is the phenomenon of the 'blue bump'.

The relevant observations are that a single power law with about ν^{-1} extended from the infrared to ~ 1000 Å has excess emission superposed on it in the range 2000 Å $\leq \lambda \leq$ 4000 Å (Grandi 1982). This excess peaks at ~ 2500 Å, so it is tempting to attribute it to some optically thick thermal source with a characteristic temperature of order 10^4 K. Such a source could be, for example, optically thick Balmer continuum. The most intriguing explanation is that the blue bump is black-body continuum from a hot accretion disk (Malkan 1983). If so, it is the only direct observation of this disk, even though the disk is a necessary part of any quasar models involving massive black holes. The jury is not in (Puetter *et al.* 1982). Many more observational consistencies can be sought, requiring improved but obtainable data. The bump can be correlated with other spectral properties, such as emission line widths, which might be expected to correlate in some way with the properties of the accretion disk. Variability in the bump, if it is really thermal, must be representative of an isotropic source, so it must agree with the scale sizes required for accretion disks. Changes in the bump would represent changes in the feeding of the disk. Improved deconvolutions of the bump from the underlying power law continuum are needed, and all bump properties need to be correlated with quasar luminosity. Hopefully, the lesson is clear. A relatively simple hypothesis stimulates the next round of data gathering. Even if the original suggestion is not confirmed, we have a head start on the next step.

So far, there is a blind spot in direct observations of quasar continua. That is the regime known as far-ultraviolet, meaning $10 \text{ Å} \lesssim \lambda \lesssim 1000 \text{ Å}$. The reason for this is partly nature's fault and partly technology's fault. Nature provided a lot of absorbing material, hydrogen at the longer wavelengths in this regime and heavier elements in the shorter, both intrinsic to the quasars and within the galactic disk through which we must observe. What radiation does penetrate is difficult to detect because conventional telescope mirror coatings do not reflect at these wavelengths, but the photons are sufficiently low in energy that the highly efficient photon counting techniques of X-ray astronomy do not work. Efforts are being made, but even the next generation of spacecraft cannot be counted upon to tell us much about quasar spectra in these wavelengths. Some characteristics can be deduced without actually seeing the continuum, by noting emission lines from ions with ionization from photons at short wavelengths. The most ionized feature so far definitely observed is from Fe xiv (Osterbrock 1981), which cannot arise without photons of wavelength 34 Å to ionize the iron. Numerous quasars have Fe x emission (Penston *et al.* 1984), requiring 53 Å photons. Even if it cannot be seen, therefore, the ionizing continuum must encompass these wavelengths.

Because of the strong hydrogen emission lines, it might be suspected that most of the continuum shortward of the Lyman limit would be absorbed by gas. This actually happens only in a few quasars (about 10%) (Smith *et al.* 1981). This is only one of several independent indications that the continuum source is not completely covered up by the gas clouds of the broad-line region (BLR; see Chapter 8) within which the continuum is immersed. Even though there are indications of gas and dust absorption associated with these clouds, such absorption does not extinct the continua. That ultraviolet and X-ray continuum can be seen to vary rapidly is evidence that many continuum photons, presumably from beamed non-thermal sources, escape unhindered through gaps between the clouds. Furthermore, at least half of the known superluminal sources are within quasars with measurable BLR. Nevertheless, the many differences between blazars and other, less variable, quasars are evidence that the presence of a BLR tends to make it more difficult to see the raw, unprocessed continuum. There are various reasons for suspecting containment of the plasma beams having relativistic bulk motion by the BLR. Seyfert 2 galaxies, with weak or no BLR, are systematically stronger radio sources (Meurs & Wilson 1984). The radio quiet quasars have strong Fe II emission features, explainable by heating from plasma beams being absorbed by the BLR clouds. Radio synchrotron continuum can also be absorbed by these clouds (Condon *et al.* 1981), accounting for why the quasars are radio quiet.

7.8   X-ray continuum

Heavy element absorption extincts X-rays at wavelengths longward of 6 Å (energies below 2 keV) in most quasars for which spectra are available. For solar abundances of the heavy elements, corresponding column densities of hydrogen of 10^{22} cm^{-2} are adequate to absorb these X-rays. The absorption is produced by K-shell electrons in the heavy elements such as Fe,Si, S and Mg. Just a single cloud of density 10^9 cm^{-3} and thickness 10^{13} cm could stop the X-rays. Fortunately, this absorption edge is fairly well defined, and X-rays shortward of 6 Å are little affected. X-ray spectra of quasars, primarily the low luminosity examples in nearby galactic nuclei, exist for 25 to 0.1 Å (0.5 to 100 keV). Shortward in wavelength of the absorption, these spectra are remarkably uniform for the quasars which have been measured (Mushotzky 1982, Petre *et al.* 1984). In fact, the X-ray spectra as a class are more homogeneous than the continuous spectra in any other wavelength regime. These X-ray continua are power law with index −0.7, with very small scatter about this index (Worrall & Marshall 1984, Reichert *et al.* 1985). The observations are primarily of low luminosity quasars, usually Seyfert 1 galaxies. There is no systematic knowledge of the X-ray spectra of high redshift, high luminosity quasars.

This X-ray spectral index is much flatter than the index required to connect the X-ray and ultraviolet fluxes. The α_{ox} (2500 Å to 2 keV) averages −1.3 (Zamorani *et al.* 1981, Gioia *et al.* 1984). Even though the X-ray spectrum has a consistent shape, this cannot extrapolate to longer wavelengths. This means that different mechanisms are responsible for the X-ray and ultraviolet continuous spectrum; in part, this may reflect the importance of the blue bump in the ultraviolet.

The interpretation of the power law X-ray continuum is that it is non-thermal. That still leaves at least two alternatives: either the X-rays are primary synchrotron radiation; or they are SSC from longer wavelength photons. The coincidence of the X-ray slope with that for optically thin synchrotron radiation in the radio is primary evidence for this non-thermal interpretation. Accelerating mechanisms must be producing an electron energy distribution appropriate for the radio synchrotron, and the same distribution accounts, either directly or by SSC, for the X-ray.

A crucial question is whether any properties of quasar continua change systematically with luminosity. If not, this constrains models for the basic mechanism which powers quasars to those capable of scaling up their generated power without using different processes to transform other forms of energy into radiation. It is only in the X-ray band that any indication of such a systematic change has been found. Various analyses show that the X-ray luminosity in the 0.5–4 keV band, L_x, does not scale linearly with the luminosity L_{opt} at optical wavelengths. Instead, $L_x \propto L_{opt}^{0.7}$ (Avni & Tananbaum 1982, Kriss & Canizares 1985). Another way to state this result is that α_{ox} would become steeper with higher luminosity. Because high luminosity quasars are also those with higher redshift, it is not definitely established that this represents a real redistribution of energy. It could be stating a difference in the form of the X-ray spectrum seen for quasars at high redshift. Alternatively, there may be systematic errors in the derived optical luminosities that are a function of luminosity, such as less dust obscuration at high luminosity or insufficient allowance for galaxy starlight at low luminosities. Regardless of the caveats, the result is important and suggestive – the kind of thing observers have to seek – especially as it can be explained by certain accretion disk models (Tucker 1983).

There is a pressing reason other than the nature of quasars for needing to describe their spectra. When the X-ray sky is observed, a diffuse background filling the entire sky is brighter than the integrated light from known individual sources. Only at millimeter wavelengths (where the 2.7 K cosmic microwave background dominates) does this happen anywhere else in the spectrum. It is truly an extraordinary result of X-ray astronomy and one of its most basic puzzles (Boldt 1981). This X-ray background is a puzzle not so much because of its intensity, but because of its spectrum. Simply taking the expected broad band contribution of all of the quasars that should be there according to optical counts can explain a substantial portion of the background. Incorporating quasar evolution could account for most of the rest, except for one fundamental anomaly. The X-ray background shows a well defined spectrum with thermal shape, corresponding to temperature $\sim 10^8$ K. This spectrum is different from the power law quasar spectra, primarily in that it is much flatter. If all quasar spectra have the same -0.7 power law already described, no combination of quasars with or without evolution can produce the spectrum of the X-ray background. To synthesize this background with quasars would require them to have power law spectral indices ~ −0.4 (De Zotti *et al.* 1982). Modelling, arguing, diddling evolution parameters cannot help resolve this issue; it cannot be resolved until we know whether the faint quasars that dominate optically derived quasar counts really have spectra like the nearby quasars which have been measured. That answer must wait upon the availability of AXAF.

Observations of the X-ray continuum are particularly influenced by variability of the quasars. Quasars are variable on short time scales (less than years) but not variable, in general, on time scales as short as would be expected for material in the immediate vicinity of an MBH (Tennant & Mushotzky 1983). Adequate instrumentation is not in orbit for careful monitoring of the X-ray spectrum in quasars. Models of the central quasar engine are severely constrained by the results of X-ray variability. Detailing this will be another eagerly awaited chore for AXAF, especially in determining spectral variability properties as a function of luminosity. It is imperative to measure the spectra well enough to learn if the variable component is non-thermal or not (Halpern 1985). The MBH models with accretion disks place very hot gas near the hole, whose thermal variability would be an isotropic luminosity change determined by the accretion rate. But non-thermal continua, particularly X-ray synchrotron, could arise from beaming. This could arise from any ejection of relativistic particles, and the time scale would not have to relate to events in the vicinity of the MBH.

7.9   Bolometric luminosities

Even though there remain disputes and uncertainties concerning the real nature of quasar continua, the overriding observational result is the similarity from object to object (Kriss & Canizares 1982). It is convenient, therefore, to deduce some simple scaling laws for estimating quasar luminosities based upon luminosities observed in restricted bandpasses. From the data already reviewed, a typical quasar continuous spectrum is taken to be a composite of power laws, with slopes of: $(a) -1$ from the optical wavelength of 5000 Å to the limit of the spectrum in the far infrared, adopted as 200 μm; $(b) -0.5$ from 5000 to 1000 Å, close to the Lyman limit at which absorption by hydrogen can substantially diminish the observed continuum; $(c) -1.3$ from 1000 to 10 Å, to agree with the empirically determined values of α_{ox}; $(d) -0.7$ from 10 Å (1.2 keV) to 0.12 Å (100 keV) from the observed shape of the X-ray spectra. The total quasar luminosity, L_{bol} in erg s^{-1}, is taken as the luminosity radiated in the continuous spectrum between the limits of 100 keV and 200 μm. Then, the following relations hold: $L_{bol} = 2.3 \times 10^{14} L_{IR}$, for L_{IR} the 25 μm luminosity in erg s^{-1} Hz^{-1}; $L_{bol} = \text{dec}(36.78 - 0.4\,M)$, for M the absolute magnitude; and $L_{bol} = 1.4 \times 10^{19} L_x$, for L_x the

2 keV luminosity in erg s^{-1} Hz^{-1}. Using the empirical result between continuum luminosity and Hα luminosity from Chapter 5, there is also the result $L_{bol} = 10^{2.9} L (H\alpha)$. For the spectrum described, 31% of the luminosity is in the optical–infrared band (a), 13% in the ultraviolet–optical band (b), 29% in the X-ray–ultraviolet band (c), and 27% in the X-ray band (d).

7.10   Radio continuum

Quasars whose continuum properties at all other parts of the spectrum appear otherwise identical can have drastically different radio powers. Less than 1% of discoverable quasars have radio fluxes as strong relative to the optical as the classical 3C and 4C quasars which provided the first quasar samples. We now know that radio bright quasars are very unusual. The astrophysical question is whether these differ in any fundamental way from other quasars, or whether the radio flux is a minor detail whose comings and goings are pretty much irrelevant to the quasar phenomenon. Observationally, the first step is to determine a distribution function which describes the fraction of quasars which are radio bright, weak, or in between. Such a function could be defined in terms of a radio–optical spectral index, analogous to the indices describing connections between other spectral regions. If a quasar had an optical B magnitude of 18, for example, and a ν^{-1} spectrum that continued to radio wavelengths, the 6 cm radio flux would be 38 Jy. This compares with the strong sources in the first radio catalogs, most of which had optical identifications in the 18–20 mag range.

Most quasars of 18 mag now known are much weaker radio sources, the majority being fainter than 1 mJy. The optical–radio spectral index for one of these is comparable to zero. Relations between radio and optical flux were described initially in terms of the ratio of fluxes rather than an effective connecting power law index (Schmidt 1970). A distribution function in either integral form, $G(>r)$, or differential form, $\Psi(R)$, describes the fraction of quasars with differing radio–optical flux ratio R. At the present time, R is most commonly taken as the ratio of radio flux at 6 cm to optical flux at 2500 Å (Wills & Lynds 1978). The former is chosen because it is a common and efficient observing band for radio telescopes (5000 MHz); the latter is a representative optically observed wavelength for quasars with moderate redshifts. It is desired that these calibrating wavelengths be in the quasar rest frame, so it is necessary to translate from observed to intrinsic wavelength by adopting a spectral index. Most quasars detectable as radio sources are compact, flat-spectrum sources with $\alpha \approx -0.5$, the same as typical for the optical–ultraviolet index.

The flux in the quasar rest frame at 6 cm, f_{5000}, relates to observed flux S_ν by $\log f_{5000} = -26 + \log S_\nu + \alpha \log (5000/\nu) - (1 + \alpha) \log (1 + z)$. ν is the observer's frequency in MHz (ν_0 in the notation of Chapter 3), S_ν is in mJy, and f_{5000} is in erg cm^{-2} s^{-1} Hz^{-1}, for comparison with $f(2500$ Å$)$. The $f(2500$ Å$) = f(4400$ Å$)(0.57)^{-\alpha}$; using the conversion of B mag to flux in equation 3.15a, $\log f(2500) = -0.4B - 19.36 + 0.24\alpha$. Taking $\alpha = -0.5$ for both spectral regions, these results yield that $\log R = -6.52 + 0.4B + \log S_\nu - 0.5 \log (5000/\nu)$.

Several attempts have been made to detect the radio flux from optically discovered quasars. Using the detections and limits in Sramek & Weedman (1980), Condon *et al.* (1981) and Smith & Wright (1980), there are 35 detections of 344 quasars. The $G(>R)$ resulting from these is shown in Figure 7.1. This displays the fraction of quasars, $G(>R)$, having radio to optical flux exceeding R. For calibration, note that an 18 mag quasar with a 6 cm flux of 1 mJy would have $R = 0.68$, at which $G(>R)$ is 20%. The curve demonstrates the small fraction of quasars having large R. The faintest values of R to which it will be feasible to extend this curve would correspond to the brightest optical quasars, $B \approx 16$, observed to the faintest possible

Figure 7.1. Function $G(>R)$ deduced from radio observations of optically discovered quasars. Horizontal bars: differential values for quasars detected within each bin of $\log R$, with numbers giving the actual number of quasars detected within the bin. Curve: integrated values of the differential detections, which is the curve of $G(>R)$.

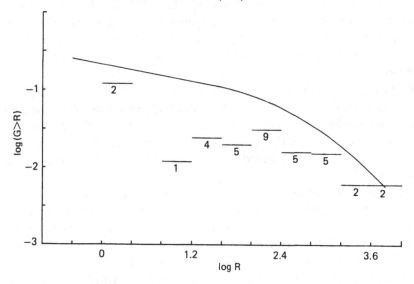

radio levels, $S_\nu \approx 0.05$ mJy, which is $R = -1.4$. At such levels of radio flux, the sky is very confusing (Condon 1984, Windhorst *et al.* 1985), because many faint galaxies (probably similar to the starburst galaxies described below) begin to appear. In fact, it could be suspected that at $R < -1.5$ most galaxies associated with quasar nuclei would begin to show. In this case, the detection is meaningless as far as it relates to the quasar properties.

7.11   Thermal vs. compact radio sources

Because of the high spatial resolutions achievable, thermal and non-thermal sources can be observationally distinguished in radio obser- vations even with single frequency observations. The two categories differ greatly in surface brightness. Classically, radio flux measurements were described in terms of the source 'brightness temperature' T_b. This is a measurement of the surface brightness translated from flux to temperature units. The concept of T_b arises because radio observations are always at sufficiently long wavelengths for observed thermal continuum sources to be far beyond the black-body maximum. In this case, the specific intensity of the source $B_\nu = 2kT\lambda^{-2}$. That is, specific intensity is directly proportional to temperature at a given observing wavelength. Radio observations measure total flux from a source, S_ν. To deduce specific intensity, which has units of flux per unit solid angle, the S_ν has to be divided by the angular area of the source. If the source has radius θ in radians, the specific intensity is simply $S_\nu/\pi\theta^2 = 2kT\lambda^{-2}$. Converting units to measure angular radii in arcseconds, λ in cm, and flux S_ν in mJy, this yields $T_b = 0.5\,S_\nu\,\lambda^2\theta^{-2}$. This source brightness temperature must be distinguished from the observed antenna temperature in the cases where the source is not resolved. Then, the relevant angular size that yields the antenna temperature is the angular resolution (beam size) of the telescope. The antenna temperature observed scales to the brightness temperature by (source size/telescope spatial resolution)^2. This also accounts for the confusion between the terms flux and flux density as used by radio astronomers. Any radio observation is actually a measure of flux emanating from the area of the observing beam, or flux density. As long as the source is unresolved (smaller than the beam), this is the same as the source flux.

For thermal sources T_b can never exceed the actual source temperature, and it equals it only in the optically thick sources. Few radio thermal sources are optically thick. For examples of thermal radio sources radiating free–free emission at $T = 10^4$ K, typical 6 cm fluxes might be 1 mJy for sources of radii $\theta = 1''$. This translates to a T_b of 18 K. That the result is much less than the temperature of the radiating gas is because free–free emission is generally optically thin. Consequently, thermal sources would never yield T_b exceeding 10^4 K. Any source observed with a brightness temperature above this must be non-thermal.

The brightness temperatures of non-thermal sources can be extremely high. To get an idea of how they might scale, realize that the electron velocities involved are relativistic, so are close to 3×10^{10} cm s^{-1}. For temperatures of 10^4 K, typical velocities are 7×10^7 cm s^{-1} because $mv^2/2 = 3kT/2$. For T scaling with v^2, the analogous 'temperature' of a non-thermal source with relativistic electrons would be $\sim 10^{10}$ K. Using more precise theoretical arguments and empirical results, it is found that T_b for optically thick (self-absorbing) synchrotron sources is about 10^{12} K. Using this gives a specific relation between source size and flux (independent of source distance) for a non-thermal source. This relation is $\theta'' = (5 \times 10^{-13} \, S_\nu \lambda^2)^{0.5}$. It is useful to note that another indication of such a source is that θ increases in proportion to λ.

Making this relation precise for high redshifts would require inserting the several cosmological terms from Chapter 3. Knowledge of the spectral shape would be needed to transform the S_ν in the observer's frame to that in the source frame. In all respects, the final result is analogous to the surface brightness calculations described in Chapter 4. It was seen there that the surface brightness (in units using Hz^{-1}) diminishes with $(1 + z)^3$. Because the measured T_b is equivalent to a surface brightness, these terms should also enter. This means that actual brightness temperature in the source frame is $(1 + z)^3$ times the observed brightness temperature. Because the numerical value in this relation is uncertain because of other theoretical and observational reasons, and because observations have generally been of sources at low z, the complete cosmological formulation rarely appears, and the result given is acceptable for most uses.

The technique described for estimating source size is important because of the availability of VLBI techniques for mapping sources with extremely high spatial resolutions. An estimate can be reached of whether a compact source should be resolvable or not. If the source was smaller than the expectation, it would mean that a source had been located with exceptional properties. If it was larger, it would show the distribution of various independent components; that is, the source structure would be resolved. Resolving compact, non-thermal radio sources is the *only* way to obtain a direct measurement for the size of any continuum-generating region in a quasar. This explains why the potential of systematic mapping with the VLBA is of such crucial significance to determining the structure in the vicinity of quasars' central engines.

7.12   Starburst nuclei

Because of the wide variety of radiation mechanisms which can take place in quasars, it is easy to feel dismayed about prospects for a fundamental understanding. It is important to be reminded, therefore, that progress is being made in sorting out the wide variety of energetic objects in the universe. A good example is the success in isolating those luminous galactic nuclei which show many superficial resemblances to quasars but which prove to be powered by a fundamentally different mechanism. These are the starburst nuclei, which can compare in luminosity with quasars. Furthering the observable similarities are strong emission lines, blue optical continuum, non-thermal radio emission, strong infrared continuum and moderate X-ray continuum. In many cases, these properties are restricted to an unresolved galactic nucleus (Weedman *et al.* 1981, Balzano 1983).

There are a few fundamental differences between starburst nuclei and quasars which illustrate the major distinctions of quasars. Easiest to notice is the absence of any BLR for the emission lines. No part of the continuum is variable, so there is no evidence for relativistic beaming effects. In the ultraviolet continuum observed with IUE, absorption lines characteristic of hot, massive stars are seen. The latter observation clinches the argument that these hot stars, forming in great numbers, provide the fundamental power source. For the starburst nuclei, luminosities in all spectral regions relate consistently to this interpretation.

The term starburst describes an event whereby the formation of stars in a galaxy is proceeding at a rate far greater than could be sustained over the life of the galaxy. This star formation is often restricted to volumes small by galactic standards, so a starburst describes a brief, localized and intense episode of star formation. In the terminology of the Armenian school, where the search for such events was a primary motivation for the Markarian galaxy survey, these are called superassociations.

Because of the formation and death of massive stars, there are observable consequences throughout the continuous spectrum. X-rays are produced both by accretion onto massive, compact stellar remnants and by supernova remnants. Ultraviolet and visible continua arise from the hot stars themselves. Emission lines are produced when these stars ionize nearby gas. Any dust associated with the star-forming region is heated by the strong stellar radiation, and the dust re-radiates a strong continuum in the infrared. Finally, supernova remnants produce non-thermal radio continua.

All of these radiation mechanisms can be related to the number and properties of the massive stars involved in the starburst. The basic parameter describing star formation is the 'initial mass function', which defines the number of stars formed as a function of their mass, $dN \propto m^{-x}dm$, with x an index that controls the shape of the IMF. This index seems to be between 1.5 and 3.5 for star-forming regions; determining it is one of the objectives for studying star-formation episodes. Upper and lower mass limits also need to be specified. Complex star-formation models can be required, because the observed mass function is affected by the different evolution rates of stars of different mass. That is, more massive stars complete their life cycles more quickly, so the more massive stars would progressively disappear as the starburst aged. Time parameters involving the duration of the starburst and the elapsed time since the starburst are incorporated in sophisticated modelling (Huchra 1977, Rieke *et al.* 1985).

The simplest circumstances arise when it can be assumed than an ongoing starburst is observed, such that dying massive stars are being replaced by newly formed ones. This is a reasonable assumption for those objects with the most intense starbursts (Gehrz *et al.* 1983). Under these circumstances, the starburst is described completely by the IMF and the duration time T over which the starburst continues. The IMF is given by $dN = N_0 m^{-x}dm$, between lower mass limit $m(\mathrm{l})$ and upper mass limit $m(\mathrm{u})$. N_0 has units of number y^{-1} so that dN is the number of stars forming y^{-1} within mass interval dm.

Stars are forming and stars are dying within a starburst. At any given elapsed time t after the starburst began, the total number formed within a mass interval is $N_0 m^{-x}t\, dm$. A fraction of these have already died by time t, the fraction being $1 - \tau(m)t^{-1}$ of the number formed, where $\tau(m)$ is the lifetime of a star of mass m. As a result, the number of stars actually visible at any time within interval dm is $N_0 m^{-x}\tau(m)dm$. That is, the number visible is simply the difference between the number that formed and the number that died.

Various useful relations among spectral characteristics follow from these concepts. For example, the bolometric luminosity of all thermal radiation from the starburst must be the integrated luminosity of all visible stars, or

$$L_{\mathrm{bol}} = N_0 \int_{m(\mathrm{l})}^{m(\mathrm{u})} l(m)m^{-x}\tau(m)dm.$$

This is primarily useful because most of the bolometric luminosity seems to be absorbed and re-radiated in the infrared by dust, making starburst galaxies very strong infrared sources. Measuring the infrared flux yields the closest observable approximation to L_{bol}. In this equation, $l(m)$ is the bolometric stellar luminosity as a function of mass. It is somewhat uncertain, depending on both stellar interior and atmosphere models. A tabulation is given in Table 7.1 from combining results of Stothers (1972) and Kurucz (1979).

Because starbursts are brief episodes, their characteristics are dominated by the luminous, massive stars which are short lived. It is the presence of such stars that gives the notable luminosity and ionization to starburst regions. As a result, all stars within the starburst IMF can be assumed to form supernovae. For a starburst in progress, with stars dying at the same rate other stars of the same mass are being born, the supernova rate R is the same as the birthrate, or

$$R = N_0 \int_{m(l)}^{m(u)} m^{-x} dm.$$

Some starbursts are so intense that this rate is several per year, so it could be checked observationally by monitoring the starburst. More indirectly, R can be estimated by the integrated non-thermal radio luminosity of the super-nova remnants. This can only be done if the time-averaged luminosity of a remnant is known. Using empirical data for Galactic remnants, Ulvestad (1982) estimates the relation $f_\nu(1465 \text{ MHz}) = 5.6(3.6)^{\alpha_r+0.75}(d/100)^{-2}R$. Here, f_ν (1465 MHz) is the observed non-thermal component of radio flux at 20 cm in mJy, α_r is the non-thermal radio continuum spectral index, and d is the distance in Mpc.

A final measure of particular importance is the total mass of stars cycled through the starburst. This is important because it describes both the amount of raw material needed and the mass of compact remnants left behind as dead stars. These corpses may provide important sources for eventual powering of luminosity generation by accretion. The total mass of stars formed in interval dm throughout the duration t of the starburst is

Table 7.1. *Bolometric luminosities of massive stars*

Mass $(M_\odot)$	$l(m)$ $(10^{37} \text{erg s}^{-1})$	Mass $(M_\odot)$	$l(m)$ $(10^{37} \text{erg s}^{-1})$	Mass $(M_\odot)$	$l(m)$ $(10^{37} \text{erg s}^{-1})$
6	0.5	22	21	38	67
8	1.5	24	24	40	74
10	3	26	29	42	81
12	5	28	35	44	87
14	7.5	30	40	46	94
16	10	32	45	48	101
18	15	34	53	50	108
20	17	36	60		

the mass per star times the rate of star formation per mass interval times T, or $m \, dN \, T$, so

$$M = N_0 T \int_{m(l)}^{m(u)} m^{(1-x)} \, dm = MT$$

for $\dot{M}$ the rate of mass cycling through all stars, in $M_\odot \, y^{-1}$.

The parameters L_{bol}, R, $\dot{M}$ and N_0 are compared in Table 7.2 for a few illustrative values of x, $m(l)$ and $m(u)$. Values are derived from the relations discussed plus the $\tau(m)$ from Larson (1974), given as $\log \tau(m) = 10.02 - 3.57 \log m + 0.9(\log m)^2$. Various other observable characteristics of the starburst could be defined, particularly the ultraviolet or visible continuum and luminosity of hydrogen emission lines. These characteristics are severely influenced by obscuring dust, so they are not described here. For attempts to determine the amount of dust obscuration, relations analogous to those already described can be used to compare the ultraviolet with infrared or radio fluxes.

Starburst galaxies can be very luminous, known examples having bolometric luminosities that exceed 10^{45} erg s^{-1}. Particularly in the infrared continuum, dusty starburst systems are difficult to distinguish from dusty quasars. If galaxies in their formation phase at high redshift are as characterized by dust as nearby starburst systems seem to be, infrared observations will be crucial to picking out these forming galaxies. Delineating high redshift, dust-obscured quasars from high redshift, dust-obscured young galaxies may prove to be a challenging task.

Table 7.2. *Samples of starburst parameters*

IMF parameters $m(l) - m(u)$	x	L_{bol} (erg s^{-1}) N_0 (number y^{-1})	L_{bol} (erg s^{-1}) R (number y^{-1})	L_{bol} (erg s^{-1}) $\dot{M}$ ($M_\odot$ y^{-1})
11 − 30	3.5	1.0×10^{42}	1.1×10^{45}	7×10^{43}
11 − 50	3.5	1.2×10^{42}	1.2×10^{45}	7.3×10^{43}
31 − 50	3.5	1.5×10^{41}	2.9×10^{45}	7.6×10^{43}
11 − 30	2.5	1.8×10^{43}	1.3×10^{45}	7.6×10^{43}
11 − 50	2.5	2.3×10^{43}	1.4×10^{45}	7.2×10^{43}
31 − 50	2.5	6.0×10^{42}	3.0×10^{45}	7.9×10^{43}
11 − 30	1.5	3.1×10^{44}	1.3×10^{45}	7.2×10^{43}
11 − 50	1.5	5.4×10^{44}	1.7×10^{45}	7.2×10^{43}
31 − 50	1.5	2.4×10^{44}	3.1×10^{45}	8.0×10^{43}

8

EMISSION LINE SPECTROSCOPY

8.1   Introduction

Astronomy in the early part of this century demonstrated that galaxies were systems made of tremendous numbers of stars. Spectroscopy of galaxies revealed the absorption lines that would be expected in the composite light from stars of different spectral classes. Galaxies showing dominant emission lines in their spectra were recognized as highly unusual. The first of these to be studied, NGC 1068, was commented upon even before the real size and nature of galaxies were understood (Slipher 1918). For several decades, because of their rarity, such galaxies were sufficiently outside mainstream research as to be given little attention. The subject of emission line spectroscopy for extragalactic objects suddenly became extremely important with the discovery of quasars, whose visible spectra are characterized by strong emission lines. Emission lines can provide diagnostics of velocities, temperatures and densities unavailable from any other technique. The lines which can be seen represent a wide range of ionization, so line fluxes also provide indirect measurements of unobserved portions of the continuum. Not least is the fact that emission lines are spectroscopically conspicuous, calling attention to locations where unusual events are occurring. The general similarities among the emission line spectra of quasars, and the scaling of these lines with the continuum source, means that the emission line spectrum is a characteristic quasar feature. To understand the origin of these lines, it is necessary to review the general physical concepts of spectroscopy.

The volumes within which quasars generate their extraordinary luminosities are volumes characterized by hot, highly ionized gases into which energy is input by many mechanisms. These gases are inhomogeneous, with complex temperature and density structure. Within the densest regions – those closest to the quasar's central engine – the initial energy generation by the quasar occurs, probably powered by gravitational accretion into the vicinity of a compact object or objects. Some of this energy is thermalized, and some is transformed into non-thermal radiation via the kinetic energy carried by particles moving at relativistic velocities. No one pretends that the center of a quasar is a place that is easy to describe. So many things are happening that virtually any part of the spectrum carries important and unique information. X-rays arise from the regions of highest temperatures and most energetic particles, the ultraviolet and optical continuum from similar but cooler volumes, the radio and part of the infrared from volumes of low particle density but high magnetic field where the kinetic energy of relativistic charged particles is converted to synchrotron radiation. Further complicating what is observed is the reprocessing of many forms of radiation into other forms by various material in the quasar; some infrared continuum, for example, is re-radiation from cool dust of higher frequency radiation which the dust absorbed. The emission lines represent re-radiation by individual ions of energy derived from a more primary continuum. Absorption by various clouds of cooler gas affects the X-ray, ultraviolet, optical and radio spectra.

All of these mechanisms need to be reviewed, because all are important. It is necessary not to be dismayed by the overwhelming complexity of what is happening. I do not think even another two decades of work will succeed in mapping the interior of quasars in great detail. The realistic objective is to determine the overall luminosities involved, to learn which energy-generating mechanisms are primary, and to decide the nature of the compact source which fundamentally powers the quasars. These objectives require mapping the density and temperature structure, at least approximately, determining the masses and kinds of material involved, and measuring the luminosities. Doing these things requires both continuum spectroscopy and line spectroscopy. This chapter describes what we can learn from emission lines.

Of necessity and from precedent, the techniques developed for studying emission lines from gaseous nebulae (H II regions and planetary nebulae) have been applied to quasars. These techniques are thoroughly discussed by Osterbrock (1974), and only the basics will be summarized here. More important is to point out the modifications to conventional techniques that have to be kept in mind for quasars. The environment of a quasar is very different from that of an H II region or planetary nebula, so parallel analysis techniques should be used with care. Much recent research effort has gone into modifying the earlier techniques for application to quasars.

As is the case in any gaseous nebula, we deal with two fundamental categories of emission lines: those produced by transitions occurring during ionic recombination following ionization, and those produced by transitions following collisional excitations by other particles in the gas. The former, recombination lines, are most importantly the lines of hydrogen. Collisionally excited lines are usually also forbidden lines, so the topics are sometimes divided into a discussion of the hydrogen lines and the forbidden lines. This is acceptable, as long as it is realized that collisional effects can sometimes be important also for hydrogen and other permitted lines, and that a few other lines besides hydrogen are attributable to recombinations. To illustrate the most important lines actually seen in quasar emission spectra, a list is given in Table 8.1. No two quasar spectra are precisely the same; the relative fluxes listed are primarily as averaged by Baldwin (1979) with some modifications from other references in this chapter. Line widths are not displayed, because that is a complex topic to be discussed separately. This simple list could be used to identify almost any quasar spectrum encountered. Fortunately, there are not a lot of strong emission lines present, so the few that are present are so characteristic that they cannot be confused with anything else. The distinctive patterns make it possible to determine line identifications unambiguously even when only two lines are seen in a spectrum. Even if no techniques of spectroscopy are understood, an astronomer can use these emission lines to determine a quasar redshift.

It is no accident that the same lines are seen in other emission line galaxies, H II regions and planetary nebulae, all often with strikingly similar relative

Table 8.1. *Quasar emission lines*

Wavelength (Å)	Ion	Relative intensity
1034	O VI	20
1216	H I (Ly α)	100
1240	N V	20
1400	Si IV + O IV	10
1549	C IV	50
1640	He II	5
1909	C III]	20
2798	Mg II	20
3426	[Ne V]	5
3727	[O II]	10
3869	[Ne III]	5
4861	H I (Hβ)	20
4959	[O III]	20
5007	[O III]	60
6562	H I (Hα)	100

fluxes to those from quasars. The reason for these similar spectra is that emission lines of the wavelengths displayed only arise within a fairly restricted temperature and density regime. Because the mix of elements in gas is similar throughout the observable universe, any place where the appropriate temperatures and densities arise will produce similar emission line spectra. What may be drastically different is the source that provides the energy for ionizing and heating the gas, but the plasma, once heated, will appear similar to an observer regardless of that source. The great advantage of this situation is that the techniques to be described are applicable to a wide variety of objects. The disadvantage is that many of the truly fundamental differences among objects are not apparent just from the relative fluxes in an emission line spectrum.

8.2   Equilibrium for hydrogen emission lines

The most significant hydrogen lines are Ly α (level 2 to level 1), Hβ (level 4 to level 2), Hα (level 3 to level 2), and Pα (level 4 to level 3). Fluxes in these lines yield various information about physical conditions in the quasar. The basic parameter describing the energy carried by an emission line is the emission coefficient j. This has units of erg cm^{-3} s^{-1}, so it measures how much energy is being released per unit volume of the gas in a given emission line. It will be written here for particular lines as $j(\text{H}\beta)$, etc., although another common notation for the same line would be j_{42}. In general, this is j_{mn} for m the upper level of the transition and n the lower. Total luminosity of a given emission line, in the absence of any obscuration, is $L = jV$, where V is the volume in which the emission occurs.

To relate the emission coefficient to the physical emission process, $j_{mn} = N_m A_{mn} h\nu_{mn}$. N_m is the number of atoms cm^{-3} with the electron in level m, A_{mn} is the transition probability in s^{-1}, and $h\nu_{mn}$ is the energy of the photon released by the transition. Both A_{mn} and $h\nu_{mn}$ are known constants for any particular transition. For hydrogen, $A_{mn} \approx 2 \times 10^{10} \, (n^{-2} - m^{-2})^{-1} n^{-3} m^{-5}$ s^{-1}. The only variable, therefore, which controls the emission in a line is the level population N_m. All of the complexities of interpreting the hydrogen emission line spectrum relate to determining N_m.

The N_m are found from a set of equilibrium equations which describe all of the transitions that can occur within the atom. The differences between hydrogen emission lines from region to region or object to object relate to the relative significance of the various terms. All terms will be given and the spectral interpretation discussed in context of those which are important. The purpose of an equilibrium equation is to describe all the possible ways of populating a level, in units of populations s^{-1}, and to equate this to all the ways of depopulating the same level. The definition of equilibrium is that the rate of populations must equal the rate of depopulations. If these were not equal, a level could become empty (if depopulations exceeded populations), and no emission lines arising from that level would be observed. Conversely, if populations exceeded depopulations, the level involved would soon collect all available electrons, so no transitions of any kind would subsequently be observed. That we observe emission lines is adequate empirical evidence that equilibrium exists.

The full equilibrium equation (ways in = ways out) for any level $n > 1$ of hydrogen is

$$N^+ N_e \overline{\sigma_{\infty n} v} + \sum_{m=n+1}^{\infty} N_m A_{mn} + \sum_{m=n+1}^{\infty} N_m J_{mn} B_{mn} + \sum_{p=2}^{n-1} N_p J_{pn} B_{pn}$$

$$\quad (1) \qquad\qquad (2) \qquad\qquad\qquad (3) \qquad\qquad\qquad (4)$$

$$+ \sum_{m=n+1}^{\infty} N_m N_e \overline{\sigma_{mn} v} + \sum_{p=1}^{n-1} N_p N_e \overline{\sigma_{pn} v}$$

$$\qquad\qquad (5) \qquad\qquad\qquad (6)$$

$$= N_n \sum_{p=2}^{n-1} A_{np} + N_n \sum_{p=1}^{n-1} N_e \overline{\sigma_{np} v} + N_n \sum_{m=n+1}^{\infty} N_e \overline{\sigma_{nm} v}$$

$$\qquad\qquad (7) \qquad\qquad\qquad (8) \qquad\qquad\qquad (9)$$

$$+ N_n \sum_{m=n+1}^{\infty} J_{mn} B_{nm} + N_n \sum_{p=2}^{n-1} J_{pn} B_{np}. \qquad\qquad (8.1)$$

$$\qquad\quad (10) \qquad\qquad\qquad (11)$$

This equation is intimidating at first glance, but many terms are ignorable in practice; only a few terms dominate for any particular set of physical conditions. The terms are numbered for further explanation.

Term (1) is the population rate of level n due to recombinations from the ionized state directly into this level. Such recombinations depend upon the densities N_e and N^+ of the free electrons and protons which are recombining to form the H I atom. The cross section for recombination into level n, $\sigma_{\infty n}$, is a function of the relative velocity v between electrons and protons. $\overline{\sigma_{\infty n} v}$ denotes a weighted integral over the relative velocity distribution, although it is often an adequate approximation to adopt a $\sigma_{\infty n}$ and v corresponding to the average electron velocity at a given kinetic temperature. This term is usually significant. Because the units of $\sigma_{\infty n}$ are cm^2, v is in cm s^{-1}, and N^+, N_e in cm^{-3}, the product has the required units of number cm^{-3} s^{-1} for the rate of transitions.

Term (2) allows for radiative transitions into n from all higher levels m. This is also usually important in all H II regions. That is not the case for terms (3) and (4). One of these is significant in the circumstance that the gas is very close to a strong source of radiation. This can differentiate quasars from other, simpler H II regions. The J represent the mean intensities of photons at the relevant energies. The B_{mn} in term (3) is the coefficient for stimulated emission causing a transition from a higher level m to n. This is always ignored. The B_{pn} of term (4) is the coefficient for radiative absorption, causing a transition from a lower level p to n.

Under some circumstances, (4) is important in quasars, but not using the J from the continuous spectrum source. The continuum is ignored as a source of J. Term (4) is important when the gas is so dense as to be optically thick to its *own* photons arising for transitions to level $n > 1$. All H II regions that are visible must be sufficiently dense to re-absorb their own Lyman photons (transitions to $n = 1$). The transfer of Lyman photons through a nebula is discussed more in Section 8.3, because such photons are re-absorbed and re-emitted many times as they progress through the nebula. Transitions from level 1 to level n are not counted in term (4) nor are they included as a way from level n to level 1 in term (7). Such transitions automatically balance, so are omitted from the equilibrium equation. (This is the situation which is meant by 'case B', when that terminology is used. It refers to a nebula optically thick to Lyman photons.) For quasars, the issue is whether the H II region is also optically thick to Balmer photons. It is the possibility of Balmer photon self-absorption that is reflected in the inclusion of term (4).

Terms (5) and (6) allow for effects of collisional excitation or de-excitation by interactions with the free electrons. (Other atoms or protons do not move fast enough compared with the electrons to be significant collision sources.) The σ_{mn} are for collisional de-excitations from higher levels m into n; the σ_{pn} are for collisional excitations from lower levels. These terms were never considered important in ordinary H II regions or planetary nebulae. Term (6) may be quite important in some high density regions within quasars, primarily as it involves level 1. Term (5) can be ignored.

Having accounted for mechanisms to populate levels, the depopulations on the right-hand side of equation 8.1 can now be considered. Terms are very analogous to those on the left-hand side. Term (7) is important and does not count transitions to $n = 1$ because of the Lyman photon resonance transfer already mentioned. Terms (8) and (9) are negligible compared with term (7) because the A_{np} are large; these are highly permitted transitions. What that means is that any collisional excitation into level n from level 1 via term (6) will result in a radiative transition out of level n. This is why emission from level n is enhanced by collisional excitation effects. Also, terms (10) and (11) are ignorable because neither involves level 1.

It is noted that equation 8.1 does not apply to level 1. The equilibrium of level 1 is controlled primarily by the ionization equilibrium of the hydrogen.

Because of the short lifetimes of excited states, virtually all H I at a given time is in level 1. When ionization occurs, it is taken to occur from level 1; ionization was not included as a way out of level n in equation 8.1. The equilibrium equation for level 1 equates the rate of ionizations to the rate of recombinations for the atom; this is discussed briefly in a later section.

8.3   Relative hydrogen line intensities

Consider for now the set of equilibrium equations for all $n > 1$. They are insufficient to solve for the absolute number of atoms cm^{-3} in a given level n, but the ratio of any two levels can be found. This is very useful, and considering the result allows further emphasis on the differences between the conditions in quasars compared with those in more conventional H II regions. It is important to understand why the classical techniques developed for emission line spectroscopy do not always work well for quasars.

The hydrogen lines most usable as diagnostics are Hα and Hβ. In the absence of obscuration, their flux ratio is given by $f(\text{H}\alpha)/f(\text{H}\beta) = N_3 A_{32} h\nu_\alpha / N_4 A_{42} h\nu_\beta$. Only the ratio N_3/N_4 is an unknown. This can be solved for with the set of equilibrium equations from equation 8.1. Under the approximations of case B, the common situation for classical H II regions, all terms except (1), (2) and (7) are ignored when setting up the equations. Solving for the equilibrium level populations at temperature 10^4 K yields an expected decrement $f(\text{H}\alpha)/f(\text{H}\beta) = 2.87$. The energy seen in H$\alpha$ should be 2.87 times as intense as that of Hβ. Observing the ratio in real H II regions is a test of whether the case B approximations are valid. For quasars, $f(\text{H}\alpha)/f(\text{H}\beta)$ is usually observed to exceed 2.87, sometimes by very much. That is, the decrement is 'too steep'.

If $f(\text{H}\alpha)/f(\text{H}\beta)$ is not 2.87, something not included in the approximation must be affecting either the level populations or the flux emerging from the gas. There are several possible effects which can be difficult to disentangle. One realistic possibility is that some of the photons generated within the gas are absorbed by dust before they escape to the observer. Absorption by interstellar dust is known to be selective; blue photons are absorbed more easily than red photons. Consequently, dust absorption preferentially extincts Hβ compared to Hα, making the Balmer decrement appear larger. In the absence of any other information or assumptions, it is not possible to determine whether the decrement is set by dust absorption effects or by level population effects. One check is to observe another hydrogen line. Knowing the wavelength dependence of the dust absorption, the observed and expected ratios of this line to Hα and Hβ can be considered to see if it also deviates as expected by dust absorption. The line that has been used often for this test is Hγ, but it is sufficiently weak and close enough to Hβ in wavelength that testing the decrement with Hγ often gives inconclusive results. A better check is the Pα line, because it arises from the same level as Hβ, so the Pα/Hβ ratio cannot be dependent on the level populations and can be affected only by dust absorption. In the absence of dust, Pα/Hβ should be 0.45. Unfortunately, Pα has such a long rest wavelength (1.87 μm) that observing it requires infrared detectors. Few such observations exist, therefore.

Most corrections for dust absorption in quasars and Seyfert galaxies have been made using only Hα and Hβ. If it is assumed that deviations from the expected decrement are caused by dust, the total amount of dust absorption can be estimated. This requires adopting a form for the dust absorption as a function of wavelength, taken from observations of the way interstellar dust affects stellar spectra in our Galaxy, and assuming that the dust cloud simply obscures the emitting region rather than being mixed with it. (If the dust is mixed, the problem is no longer one simply of absorption, but scattering effects also have to be considered, as described by Mathis (1983)). For such a reddening law, there is a relation between the total amount of extinction and the amount by which the light is reddened, the latter being a measure of the differential extinction between any two wavelengths. Reddening laws are not the same in different directions within our own Galaxy. What to assume for quasars is questionable; for the correction that follows, I adopt the result in Savage & Mathis (1979). With these conventional assumptions, the correction to the observed flux of Hβ which has to be applied to determine the flux that would have been present without dust absorption is DC(Hβ) = (0.35 Hα/Hβ)$^{3.1}$. This is a multiplicative correction; the dust-corrected Hβ flux, which is necessary to derive the intrinsic Hβ luminosity generated by the source, is DC(Hβ) times the observed flux. In this correction formula, Hα/Hβ is the observed decrement.

It can be seen from this relation that the dust correction can be significant at Hβ. Typical observed ratios are about 5, giving a correction factor of 5.7. This factor is so large that the approximations made in deriving it may be significant sources of uncertainty in deducing the real Hβ luminosity. In fact, this question of the proper correction for dust usually is the greatest source of uncertainty in emission line luminosities, outweighing errors in the measurements or in the derived distances to the objects considered. Measuring redder lines is a big help for this problem, because these lines are less affected by the dust and so by the uncertainties associated with it. For example, the dust correction to Hα by the same precepts is DC(Hα) = (0.35 Hα/Hβ)$^{2.1}$. For an observed decrement of 5, the correction to Hα is a factor of 3.2. It is clear that dust corrections become less significant at redder wavelengths. Making corrections for lines at bluer wavelengths than Hβ is a very chancy affair indeed, so fluxes from such lines have to be interpreted with great reservations.

In all H II regions and planetary nebulae, in many galactic nuclei, and even in the narrow-line regions (NLR) of Seyfert galaxies and quasars, departures of the Balmer decrement from the predicted 2.87 can be safely attributed to the effects of dust absorption. Corrections for this absorption, although approximate and somewhat arbitrary, can be attempted as described above. Under the conditions of high density found in the BLR of quasars, however, other effects may steepen the Balmer decrement. These are changes produced in the N_3/N_4 ratio compared with the ratio deduced from the case B equilibrium equations. In the dense clouds that make up the BLR, disentangling effects of dust from the effects of the other mechanisms to be mentioned remains unresolved. Heroic efforts have been made to solve the problem (Lacy *et al.* 1982), but it has not been possible to reach a unique conclusion (Rudy & Puetter 1982).

If the population of level 3 is enhanced compared with that of level 4, the Hα/Hβ ratio will exceed 2.87. Observed deviations are always in this direction; decrements are never observed with the ratio less than 2.87. Any physical effects which we must worry about, therefore, are those that tend to populate lower levels preferentially compared with higher ones. The two significant effects in quasars are collisional excitation and self-absorption of the Balmer lines: terms (4) and (6) in equation 8.1.

Significant collisional excitation by free electrons, term (6), can occur from the ground state if $N_e \gtrsim 10^8$; the energetics of the collisions are such that lower levels are selectively excited compared with the higher; this enhances N_3 compared with N_4. Collisional de-excitations also occur in this situation, but when radiative de-excitations follow collisional excitations, the Hα/Hβ is enhanced compared with the case where collisional effects are ignored. Self-absorption of Balmer photons, term (4), occurs if the gas is optically thick to these photons; that is, if enough atoms are in level 2 that radiative transitions to level 2 can be re-absorbed. The overpopulation of level 2 could come about by collisional excitation, or because there is such a high density of recombining atoms that the equilibrium leaves a significant density of atoms in this state. Whatever the reason, all Balmer photons do not then escape to be observed, with some being absorbed to produce additional excitations. Following the course of Balmer photons then becomes a difficult exercise in radiative transfer (Canfield & Puetter 1981, Kwan & Krolik 1981). Because any re-absorbed higher Balmer photon has another chance to decay via a Paschen transition to level 3, followed by Hα, the net effect of self-absorption is to enhance the population of level 3 and the resulting production of Hα photons. The result is an increased Hα/Hβ ratio.

8.4   Balmer line luminosities

Once these high density effects begin to occur, diagnostics of the gas from Balmer line fluxes are no longer straightforward. One of the most important uses of Balmer line luminosities is to determine the luminosity of the ionizing continuum, discussed below. The technique to be described only works if the gas is optically thin in the Balmer lines. Within the complex dense regions of quasars, answers derived from that technique have to be considered only as upper limits to the required amount of ionizing radiation, in situations where the high density effects are enhancing the production of Hα. It may be some comfort to note that the uncertainty introduced in this way is in the opposite direction to that introduced by the dust, dust tending to obscure both the ionizing continuum and the emission line photons it produces. The procedure I will follow is to present simple relations based on the case B equilibrium, with occasional reminders of the direction of uncertainty in more realistic circumstances.

Under conditions in which the hydrogen emission lines arise strictly from decays following recombinations, the luminosity in these lines must relate to the luminosity of the continuous spectrum which is ionizing the gas. The reasoning which describes this relation is sometimes called the Zanstra method. It is important because the luminosity of the hydrogen line emission may be the only way to measure the ionizing radiation being produced by the quasar, especially if the ultimate source of this continuum radiation is shielded by gas. A nebula which contains enough gas to absorb all of the ionizing continuum impinging upon it is said to be 'ionization bounded'. Only in this case does the ionizing continuum deduced from hydrogen line luminosity represent all of this continuum; if any continuum leaks out between absorbing clouds, the values given by the Zanstra method are only lower limits to the continuum luminosity.

The reason this technique works is that the gas is optically thick to the Lyman lines, but optically thin to the Balmer lines. A Ly α photon transfers through the nebula by being absorbed, re-emitted, absorbed, etc., in successive radiative excitations and de-excitations, until the photon finally escapes from the gas or is destroyed by interaction with dust. If the photon is a higher energy Lyman photon, it will be radiatively absorbed and re-emitted until it eventually breaks down into a Ly α and a Balmer photon. For example, Ly β might be absorbed to produce an excitation to level 3, from which the following decay could be via Hα and Ly α, rather than via Ly β. Given enough chances, every Ly β photon will eventually break down this way. A little thought about all the possibilities shows that every Lyman photon eventually reduces to a Balmer photon and a Ly α photon within the gas. Even Ly α photons produced immediately following recombination are accompanied by a Balmer photon, in this case a photon of the Balmer continuum arising from recombination directly into the second level, or the Balmer photon produced by the last cascade into level 2 that preceded the emission of Ly α. Because of these processes, *each* recombination eventually leads to the production of one Balmer photon and a Ly α photon. As the recombined electron cascades down, it must produce a Balmer photon if it also produces a Ly α photon. If it decays to level 1 without passing through level 2, this produces a higher Lyman photon, which is then broken down by re-absorption into a Balmer photon and a Ly α photon. Therefore, the number of recombinations $cm^{-3} s^{-1}$ = the number of Ly α photons produced $cm^{-3} s^{-1}$ = the number of Balmer photons produced $cm^{-3} s^{-1}$.

The Balmer photons escape the nebula without absorption, so can be observed. In principle, the Ly α photons also eventually escape after many scatterings, but their longer random walk through the gas greatly increases their chances of destruction by dust. The reasoning of the Zanstra method along with the case B equilibrium predicts a ratio of the emergent Ly α flux to Balmer line fluxes. The $f(Ly\ \alpha)/f(H\alpha)$ ratio should be about 13. It is observed in quasars to be nearly unity (Wu, Boggess & Gull 1983). This great discrepancy is not hard to account for as long as some dust is mixed with the ionized gas. Given the ease with which Ly α can be destroyed by dust or absorbed by intervening H I clouds, it is perhaps surprising that enough does escape to make this the strongest quasar line. Although dust can adequately suppress the $f(Ly\ \alpha)/f(H\alpha)$ ratio, optical depth effects also do this. If the gas is thick enough for Balmer line self-absorption, this removes electrons from $n = 2$, thereby suppressing Ly α emission relative to Balmer emission. Collisional effects work in the opposite direction, however. Collisions preferentially enhance lower levels, so the N_2/N_3 population ratio would increase, thereby increasing $f(Ly\ \alpha)/f(H\alpha)$.

Because of these interpretational difficulties with Ly α, it is safer to use Balmer photons for measuring the number of recombinations. In fact, because of the impossibility of reliable dust corrections to Ly α, my suggestion for dealing with quasars in which only Ly α can be observed, because of the redshift, is to hypothesize what Hα should be on the basis of other empirically observed Ly α/Hα ratios. Then, apply the relations to be given below based on Hα.

The other concept needed to achieve the measurement of ionizing luminosity is to realize that an equilibrium must be present such that the number of recombinations $cm^{-3} s^{-1}$ = the number of ionizations $cm^{-3} s^{-1}$. The gas must retain an ionization balance. The number of ionizations is the same as the number of ionizing photons absorbed. Consequently, the number of Balmer photons emitted by the gas is equal to the number of ionizing photons absorbed by the gas. That is the basic conclusion of the Zanstra method.

This concept can be related to some useful numerical equations. Note that the equilibrium equations can be solved to determine the relative fraction of Balmer photons from any given transition compared to all Balmer photons, including continuum photons, emitted following recombination. The solution is that 45% of all Balmer photons emerge from the gas as Hα photons. Inverting this, it means that 2.2 ionizing photons have been absorbed by the gas for every Hα photon that is emitted.

While this yields the number of ionizing photons, it does not give the luminosity of the ionizing continuum in erg s^{-1} unless the shape of the ionizing spectrum is known. That is,

$$L(H\alpha)/h\nu_\alpha = \text{the number of H}\alpha \text{ photons s}^{-1}$$
$$= 0.45 \, h^{-1} \int L_\nu \, \nu^{-1} d\nu. \tag{8.2}$$

The long wavelength limit of this integral for the ionizing continuum is set at 912 Å by the ionization limit of hydrogen. Picking the short wavelength limit is more problematical. At what energy do photons cease to be used up by ionizing hydrogen? That depends on what other processes are happening in the gas. Ionizing photons can also be absorbed by other elements; a reasonable guess is to cut off hydrogen ionizations at that wavelength (228 Å) below which photons are instead used to ionize He II. This is a relatively arbitrary limit, although it is seen from the integral that for power law spectra, the number of photons rapidly decreases as their energy increases, so the higher energy photons are not in most cases important for the hydrogen ionization process. If a power law ionizing continuum of form $L_\nu = K\nu^{-1}$ is used, the luminosity normalization becomes $K = 21L(H\alpha)$. This in turn can be used to predict the ratio of $f(H\alpha)$ to the flux of the same continuum extrapolated to Hα. That is, the equivalent width of Hα would be predicted as $d\nu = f_{\nu\alpha}^{-1} f(H\alpha) = L_{\nu\alpha}^{-1} L(H\alpha)$, which in wavelength units is 312 Å. The Hα line is rarely observed to be this strong. As discussed elsewhere, such a result is used to infer that the gas does *not* absorb the entire continuum, so that the continuum is observed to be stronger than predicted relative to the lines. This implies that the ionized clouds are distributed in a way that causes them to intercept as little as 10% of the continuum.

To summarize, the procedure to determine the luminosity of the ionizing continuum in a quasar is: (a) observe the Hα flux in units of erg cm^{-2} s^{-1}; (b) use equation 3.9 to transform this to L(Hα), remembering that such a line flux is a bolometric flux measurement; (c) correct for dust absorption using Section 8.2; (d) use equation 8.2 with the corrected Hα luminosity to determine K. If only f(Ly α) is available, I recommend assuming that the Hα flux would be the same as that observed for Ly α.

Another fundamental parameter of the quasar emission region that can be determined from hydrogen line emission spectra is the mass of the emitting gas. The emission coefficient per unit volume has already been described for the Balmer lines in terms of the level populations. An alternative expression describes this emission coefficient relative to the number of recombinations that occur. For a given temperature, the recombination cross sections and relative electron–proton velocities are known for equation 8.1. As was the case for the Zanstra method, the equilibrium equations then determine what fraction of the recombinations result in production of a given Balmer line; equivalently, one could state the effective recombination cross section leading to the emission of Hα, Hβ, or any other line. Taking Hα, the result is such that the emission coefficient is j(Hα) = 3.6 $\times$ 10^{-25} $N_e N^+$ erg cm^{-3} s^{-1} at $T \approx 10^4$ K. It is conventional to set $N_e = N^+$ in volumes of ionized gas, as hydrogen is so much more abundant than all other elements that virtually all free electrons come from the ionization of hydrogen.

This is a useful numerical result. The total luminosity emitted by a volume is L(Hα) = j(Hα)V. If the density in this emitting volume is constant, the mass of ionized gas is $M = N^+ m_p V = N_e m_p V$, because $N_e = N^+$ but $m_p \gg m_e$. Rearranging and converting to standard astronomical mass units yields L(Hα) = 3.6 $\times$ 10^{-25} $N_e M m_p^{-1}$ = 4.3 $\times$ 10^{32} $N_e M$ erg s^{-1}, for M in solar masses.

This result gives a way of determining the mass of emitting gas, if the density N_e of that gas is known. Techniques for estimating N_e are described in Section 8.5. It is found that the N_e in the BLR of quasars, which is the region closest to the ionizing source, is high: $N_e \gtrsim 10^8$ cm^{-3}. Because the mass depends inversely on N_e, the mass uncertainty depends on the uncertainty in N_e. Really, we can only estimate mass limits, because N_e is so poorly known. Note also that the emission will be dominated by the volumes with high N_e because the emission coefficient depends on N_e^2. The L(Hα) of quasars range from 10^{39} to 10^{46} erg s^{-1}. These are extraordinary luminosities at the upper range. Yet, even for L(Hα) = 10^{46}, the mass of emitting gas is $M < 10^5 M_\odot$ if $N_e > 10^8$ cm^{-3}. The faintest quasars could have a detectable BLR from only a small fraction of a solar mass in ionized gas. Recall as well that some of the Hα emission may arise following collisional excitation or Balmer line self-absorption, so the emitting mass required could be even less. The point is that it does not take a very great amount of ionized gas to account for the BLR in even the most luminous quasars. This conclusion is important, because it demonstrates that this emission could be compressed into quite small volumes.

The equations relating mass and volume of the emission region can also be used to estimate the volume of that region as $V = L(H\alpha)j(H\alpha)^{-1} = 2.8 \times 10^{24} N_e^{-2} L(H\alpha) \text{ cm}^3 = 10^{-31} N_e^{-2} L(H\alpha) \text{ pc}^3$. Once again, considering $N_e > 10^8 \text{ cm}^{-3}$ and $L(H\alpha) = 10^{46}$, the emitting volume is $V < 0.1 \text{ pc}^3$. If this were spherical, it would have a radius of <0.3 pc. This shows the first indication that the regions in which quasar activity arises are very small on the scale of galaxies. Other arguments demonstrate that quasars are not evenly filled with emitting gas, but the gas is in clouds that only occupy a relatively small fraction of the total volume. As a result, the sizes of the BLR can be somewhat larger than the limits given here.

8.5   Ionization parameter

There is an independent way to judge the scale size of the ionized region, use of which requires lines in addition to hydrogen. This invokes the concept of the 'ionization parameter', which is a ratio of ionizing photon density to material particle density. It arises from the equation for ionization equilibrium between any two successive ionization states, denoted in general as N and N^+. Were we dealing with hydrogen, N would be the density of neutral hydrogen atoms in any level of excitation and N^+ the density of protons. Ionization virtually always occurs out of the ground state, however, so for hydrogen N could be taken as N_1. The equilibrium equation equates the number of recombinations s^{-1} to the number of ionizations s^{-1}: $N^+ N_e \times \overline{\sigma_r v} = N \int_{\nu_0}^{\infty} 4\pi J_\nu a_\nu (h\nu)^{-1} d\nu$. σ_r is the total cross section for recombination and a_ν is the absorption coefficient for ionization in units of cm^2 per photon. The threshold frequency which photons must have for sufficient energy for ionization is ν_0; the term $4\pi J_\nu / h\nu$ is the number of photons impinging on the unit volume containing N within a unit time. This term has units number of photons $\text{cm}^{-2} \text{s}^{-1}$. That is, this term represents a flux. Because of this, the J_ν term diminishes with the square of the distance from the source of the ionizing continuum of luminosity L_ν. For a simple geometry where the continuum source is localized at a distance d from the gas, this means that $J_\nu \propto L_\nu d^{-2}$, and d represents the scale size of ionized gas distribution about an ionizing continuum source.

This equation can be rearranged, therefore, to $N^+/N \propto L_\nu/N_e \, d^2$. The right-hand side is a measure of the ratio of photon flux to matter density. This is the ionization parameter, because it controls the ratio N^+/N. The result of interest is that the ionization ratio also depends on distance to the source. If the source luminosity is independently known, this distance can be determined from the observed ionization equilibrium.

In practice, attempting this requires a sophisticated ionization code (Davidson & Netzer 1979, Ferland & Mushotzky 1982). No single element is suitable. Only carbon has strong lines observable from different stages of ionization, C IV and C III. It is necessary, therefore, to input element abundances and solve for the ionization equilibrium of various lines.

The relations in the ionization parameter can show some simple but extremely important consequences. For example, the ratio of the C III] λ 1909 to C IV λ 1550 lines intensities varies very little with quasar luminosity, over a luminosity range exceeding 10^4 from Seyfert 1 galaxies to high redshift quasars (Wu *et al.* 1983). The observed ratio changes by no more than a factor of two, with C IV stronger in the low luminosity objects. Yet, this ratio is a measure of the ionization parameter, which depends on Ld^{-2}. This shows how the size scale of the ionized region goes with luminosity, approximately that L increases with d^2. As we might initially wonder whether the luminosity increases with source volume or with source density, this result teaches us something. It shows that as the radius increases, the luminosity does not increase in proportion to total volume, which would go as d^3. The luminosity increases by a lesser amount. This means that the line emission region tends to become more diluted as the luminosity increases. Other independent arguments confirm this; the fraction of the volume filled with emitting clouds diminishes as the luminosity increases and the volume enlarges (Mushotzky & Ferland 1984).

One would also expect from this discussion that higher ionization features would be found closer to the ionizing source. Ionizations at least up to the level of Fe x are commonly observed in quasars. Comparing the conditions shown by high compared with low ionization features indeed proves to be a useful probe of the gradient of conditions within quasars (Shuder 1982, Filippenko 1985).

8.6   The forbidden lines

Although the hydrogen lines are the most useful spectral features quantitatively, the strongest emission line in low redshift quasars is often not a hydrogen line. The 'forbidden' line of O III, [O III] λ5007, can be the strongest line in a quasar spectrum. If for no other reason than this, it is important to understand. As a group, the forbidden lines provide the diagnostics of densities and, to a more limited extent, temperatures in quasars. The forbidden lines do not arise from cascades following electron–ion recombination, as do the Balmer lines. The levels from which occur the observed radiative transitions producing forbidden lines are populated by collisional excitations from lower levels. All forbidden lines observed in quasars are collisionally excited lines.

There is no requirement that collisionally excited lines also be forbidden. It is primarily an astrophysical coincidence that this is so. The coincidence arises because of the match between the characteristic energies required for excitation and the kinetic energies of free electrons in the ionized gas. Two of the strongest permitted lines, Mg II λ2798 and C IV λ1549, are also collisionally excited. The collisionally excited lines act as the fundamental thermostat for the plasma and explain why the observed spectra and temperatures of H II regions are so similar in widely different categories of objects. Effectively, the electron gas reaches a thermal distribution that matches it to the ions available for collisional excitation. If the gas becomes too hot, collisional excitations increase, and the subsequent radiative transitions remove energy from the plasma. If the gas cools too much, excitations decrease and the kinetic energy of the electrons is retained. Because of similar elemental abundances in astrophysical plasmas, the characteristic electron temperatures are $\sim 10^4$ K, at which the equilibrium is reached between kinetic energy lost from the electrons via collisions and the radiation leaving the nebula in the collisionally excited transitions. In the absence of evidence to the contrary, it is appropriate to assume temperatures of 10^4 K for any observed emission line region. Some second-order expectations prove correct. A nebula with fewer heavy elements, for example, would have less cooling because fewer collisionally excitable transitions are available.

For determining abundances of the elements in quasars, this thermostatting mechanism is quite a disadvantage. Elements heavier than hydrogen or helium are visible only via collisionally excited lines, so the line strengths are not very sensitive to the actual abundance. As described, the temperature will rise until these cooling lines are excited to the level necessary to stabilize the temperature. As a consequence, equilibrium calculations that are sensitive to both temperature and degrees of ionization are necessary to attempt the estimation of abundances. It is unfortunate that these cannot be accurately derived, because tracing the production of heavy elements with epoch is a fundamental issue of galaxy evolution. The astonishing thing about quasars is that they seem, to within the uncertainties mentioned, to have similar abundances of carbon, nitrogen, oxygen and silicon regardless of redshift. Yet these elements had to be produced within stars and then released via stellar mass ejection before the material could be incorporated within the quasar. This means that a major episode of star formation had to predate the highest redshift quasars seen, or had to occur within the first 20% of the history of the universe. That result pushes the limits on galaxy and star formation back quite far. That so many stars must have lived and died before the quasars formed yields puzzles. The stars must have been massive to have such short lifetimes, and there must have been many of them to produce so many elements. Where are their remains today? Is there a large invisible component of galaxies formed by their corpses, neutron stars or black holes? If we are to relate quasar formation to the early phases of galaxy lives, answers must be found. A lot of help would be provided if it were possible to use quasar spectra as an accurate diagnostic of how heavy element abundances developed in the early universe. This is one motivation for striving to improve models of the emission line regions.

The analysis of forbidden lines utilizes equilibrium equations that show a critical dependence on the balance between radiative and collisional de-excitations. It is this which explains why these lines are not seen in laboratory spectra. Because the transitions are quantum mechanically 'forbidden' (meaning the A values are very low), an excited level has a long lifetime before a radiative de-excitation. If, before it radiates, the ion is collisionally de-excited, no photon leaves and nothing is seen. For this reason, the forbidden lines are very sensitive to density; if the density becomes high, the excited states are de-excited by collisions with free electrons before having time to radiate. Sufficiently low densities to allow significant radiation are not achievable in laboratory environments. It is this density effect that provides the most important quantitative use of forbidden lines for quasar interpretations.

Using the concepts of equilibrium equations, a two-level ion can illustrate the key results relating to forbidden line interpretations. When these are the only levels to consider, the equilibrium equation (rate of transitions into the level equated to the rate out) for level 2 is

$$N_1 N_e \overline{\sigma_{12} v} = N_2 A_{21} + N_2 N_e \overline{\sigma_{21} v},$$

and for level 1 is

$$N_2 A_{21} + N_2 N_e \overline{\sigma_{21} v} = N_1 N_e \overline{\sigma_{12} v}.$$

Here, the cross section σ_{12} is for collisional excitations from level 1 to level 2 and σ_{21} is for collisional de-excitations from level 2 to level 1. In concept and units, the terms are the same as discussed in Section 8.2. Any processes related to photon absorption can be safely ignored because the transition probabilities for radiative processes are so low. The perceptive reader notes that these equations are identical, so it is possible to solve only for the ratio N_2/N_1. Doing so yields the result

$$N_2/N_1 = (\overline{\sigma_{12}v}/\overline{\sigma_{21}v})[1 + (A_{21}/N_e \overline{\sigma_{21}v})]^{-1}.$$

The term that plays the most important role is the one term that is density dependent, the ratio $A_{21}/N_e \overline{\sigma_{21}v}$. What will prove to be crucial is whether this term is greater or less than unity. As a consequence, the density N_e at which $A_{21} = N_e \overline{\sigma_{21}v}$ is defined as the critical density. For densities significantly above the critical density, forbidden line emission is negligible compared to recombination lines. For densities below, the forbidden line emission is significant. Why this is the case is realized by considering the emission coefficient for the transition involved, $j_{21} = N_2 A_{21} h\nu_{21}$ erg cm^{-3} s^{-1}. With the result for N_2/N_1, this becomes

$$j_{21} = N_1 A_{21} h\nu_{21}\{(\overline{\sigma_{12}v}/\overline{\sigma_{21}v})[1 + (A_{21}/N_e\overline{\sigma_{21}v})]^{-1}\}.$$

Because N_1 is a measure of the ground state density of the ion, it is some fraction of the density of all material in the nebula and so is proportional to N_e. Quantitatively, $N_1 = X_1 N_e$, with the value of X_1 depending on the abundances and ionization equilibrium in the region of emission.

Consider first the high density case for which N_e greatly exceeds the critical density. Under this circumstance, $j_{21} \approx X_1 N_e A_{21} h\nu_{21} (\overline{\sigma_{12}v}/\overline{\sigma_{21}v})$. That is, for densities substantially above the critical density, the emission in the forbidden lines increases only *linearly* with density. Contrast that with the situation for the recombination lines and the collisionally excited permitted lines, for which emission always increases with N_e^2. This means that in a high density gas, the ratio of permitted line to forbidden line emission will increase with the density, so that the permitted lines become much stronger than the forbidden lines.

Behavior is very different in the low density case. Then $A_{21} \gg N_e \overline{\sigma_{21}v}$, and $j_{21} \approx X_1 N_e^2 \overline{\sigma_{12}v} h\nu_{21}$. The emission in the forbidden line increases with N_e^2 and continues to scale in the *same* proportion as the permitted lines for any density.

The results are very useful for quasars because they give the most direct evidence of differentiation between emission regions. In regions where the N_e is below the critical density, the ratio of forbidden lines to hydrogen line emission remains the same regardless of the actual value of N_e. This is why such a variety of objects have virtually the same intensity ratios among these lines. For the brightest forbidden lines, the [O III] $\lambda\lambda 4959,5007$ doublet, the critical density is about 10^6. The A_{21} for these two lines combined is 2.6 ×

10^{-2}. (Incidentally, this pair of lines arises from the same upper level, so the ratio of intensities is fixed by the ratio of the $A(5007)$ to $A(4959)$. That ratio is almost precisely 3. This means that the $\lambda 5007$ line is always three times as intense as the $\lambda 4959$ line. It cannot be otherwise. If it is observed to be otherwise, the only explanation is a malfunctioning spectrograph.) Contrasting to the permitted lines, the A for them are of order 10^8. This means the critical density for a permitted but collisionally excited line such as C iv $\lambda 1550$ is $\sim 10^{16}$, an extremely high density. This line, therefore, will scale with the hydrogen lines regardless of density. The most important feature sensitive to intermediate densities is the 'semi-forbidden' line of C iii] $\lambda 1909$. Its critical density is $\sim 10^{10}$.

The strength of the forbidden or semi-forbidden lines relative to the hydrogen lines or other permitted lines yields, therefore, quantitative constraints on the densities in the emission regions. These densities must be known to determine the mass estimates for the ionized gas. Unfortunately, the results for the most interesting emitting region are often only broad limits, because that region usually does not show forbidden lines. Their critical densities then provide only lower limits to the true densities. Usually, however, C iii] is seen. Densities are thereby taken to be $10^8 \lesssim N_e < 10^{10}$, as the [$O$ iii] critical density must be greatly exceeded but the C iii] not reached. Adding to the uncertainties is the fact that the emission coefficient for permitted lines goes as N_e^2, which means that the observed emission preferentially arises in the volumes of highest density. A large mass of ionized gas covering a larger volume with a lower density could be present but inconspicuous.

In terms of radiative transfer effects, the forbidden lines, the semi-forbidden lines and the collisionally excited permitted lines of C iv and Mg ii are simpler to interpret than Ly α. Once a photon is generated by one of these ions, the only process that can interrupt its escape from the nebula is absorption by dust. This applies to the permitted, collisional lines because their ionic abundances are so low that resonance absorptions are very unlikely. It applies to the forbidden and semi-forbidden features both for that reason and the reason of the very low coefficients for radiative transitions. (A forbidden emission transition is also a forbidden absorption transition.)

8.7   Emission line profiles

Because the various emission lines can have their relative intensities greatly changed by density effects, it is important to have some way of judging whether emission lines arise from the same volumes or not. For quasars, this determination is made by observing the distribution of the line emission in velocity space; that is, by observing the emission line profile. The profile describes the one-dimensional projection of the Doppler shifts caused by motions within the emitting gas. For the simplest case, in which the motions are random with a Maxwell–Boltzmann velocity distribution, the observed line intensity as a function of displacement from the line center in velocity units, Δv, is $I = I_0 \exp(-\Delta v^2/\beta^2)$. The β is the most probable velocity of a gas element in three dimensions; the most commonly measured parameter of the line is the full width at half maximum intensity, and this FWHM $= 1.66\beta$. To calibrate what is seen in quasars compared with the predominantly thermal gas motions in normal HII regions, a gas with a temperature of 10^4 K would have $\beta = 13$ km s^{-1} for the hydrogen lines, strictly from thermal line broadening.

Emission lines in quasars are orders of magnitude broader than this, which is an indication of the dramatically violent processes which must have affected the gas. Furthermore, different lines show different profiles. When profiles are the same shape, demonstrating emission at the same location in velocity space, it can be assumed that this emission is also coincident in physical space; that is, the similar lines arise in the same volume. As a result, lines with the same profiles can be taken to arise from regions of the same density. For quasars, this is a crucial diagnostic, because there are conspicuous differences in profile shapes among different emission lines.

It is the difference between the profiles of the forbidden lines and those of the permitted lines, primarily hydrogen, that is the most conspicuous spectroscopic signal of a quasar. Many other astrophysical objects have emission lines, but quasars are unique in having conspicuous differences in the line profiles between forbidden and permitted lines. This makes quasars easy to recognize, of course, and provided the fundamental classification definition for the Seyfert 1 galaxies (Khachikian & Weedman 1974). These galactic nuclei were subsequently demonstrated to have many other characteristics qualifying them as quasars, so the simple and naive classification arising from the line profiles must have some fundamental astrophysical significance. What it calls attention to is the existence of a region characterized by high densities and high velocities in the emitting gas.

The characteristic line profile for a permitted line in a quasar has two components. One component is very broad, approximately gaussian in shape, but frequently with irregularities and asymmetries, extending over a velocity range exceeding 10^4 km s^{-1}. This component is often referred to as the 'broad wings', and arises in the BLR. Superposed on this profile is another, much narrower profile which can be characterized as a separate component with width normally a few hundred km s^{-1}. This narrow profile is mimicked by the forbidden lines, so it is taken to represent a separate volume in which densities are below critical densities of the forbidden lines and in which velocities are more moderate. This is the NLR. A proper description of the ionized gas within quasars must at least distinguish between the BLR and the NLR. There is probably a continuous transition between them.

The most important differences between the regions are in their extent and mass. The profiles from the BLR have immense potential for detailed observation, because they definitely vary in shape and luminosity on time scales of order a year (Capriotti, Foltz & Peterson 1982, Ulrich *et al.* 1984). This is an extremely important result, because it demonstrates that this emission arises in volumes of size scale less than a light year. There is the excellent opportunity to learn a great deal about the distribution of gas in this region from careful monitoring of profile shapes. Interpretations of some observations are discussed in Chapter 9. Using the equations from Section 8.3, it is seen that the BLR does not require much mass to produce the observed emission. For the most luminous Hα and the knowledge that the densities must not exceed $\sim 10^{10}$ cm^{-3} critical density of C III], but must exceed substantially the critical density of [O III], the H II mass is only $\sim 10^5$ M$_{\odot}$ (for $N_e = 10^8$). The observed BLR in faint quasars require no more than 1 $M_{\odot}$. Consequently, the masses of ionized gas required in the BLR are very small compared with the masses required for accretors within scenarios where the quasar power derives from MBH or other compact objects. Some description can be made of the velocity structure of the BLR from the nature of the profile asymmetries (Osterbrock 1984). Hints that the dense clouds are showing a disk shaped distribution are an example of how this material, even though insignificant in its mass content, plays a vital role in providing diagnostics for central regions of the quasar. Similarly, ionization models of the BLR yield important, indirect information about the fundamental radiative processes involved in the quasar engine.

By contrast, the NLR must be much larger and contain far more ionized gas. In a few Seyfert 1 nuclei, this region is resolvable with observed radii of order 100 pc. Line profile variations are not observed, and the large masses require a large volume to contain them. The densities must not exceed the critical density of forbidden lines, and Hα luminosities from the NLR can be comparable with those of the BLR; for density less than 10^6, Hα luminosity of 10^{46}, the H II mass exceeds 10^7 $M_{\odot}$. Because the NLR is so much larger than the BLR, the conception of the relation between them is that the BLR is located near the ionizing continuum source, all immersed within the center of the NLR. Ionizing photons from the central source are able to penetrate to the NLR (Collin-Souffrin 1978).

Virtually by definition, having a BLR means that an object is a quasar. This criterion was important in Chapter 5 in locating the lowest luminosity examples of quasars. Various other categories of active galactic nuclei show NLR, without accompanying BLR, and it is not yet obvious whether any such categories are fundamentally similar to quasars. The recognized categories are the Seyfert 2 galaxies and the starburst nuclei. The former are a great puzzle, as the source of ionization for the NLR is not understood. The degree of ionization and kinetic energy revealed by the velocity structure within Seyfert 2 nuclei are comparable with the NLR in Seyfert 1, but there may be distinguishable differences (Cohen 1983). A great deal of effort has gone into improving spectroscopic descriptions, using refined classifications based on the relative visibility of BLR and NLR (Osterbrock 1981). This makes possible quantitative studies of the range of nuclear structure, basically to determine if Seyfert 2 nuclei have an obscured or a quiescent BLR. That is, should these nuclei also count as quasars? That Seyfert classification schemes are useful primarily as a working convention rather than descriptions of fundamentally different phenomena is implied by the observation that at least a few objects have BLR spectra so variable as to change their classification from Seyfert 1 to 2 (Penston & Perez 1984). Nevertheless, I prefer to maintain the convention that a galactic nucleus should not be called a quasar unless an observable BLR is present.

It is not possible to distinguish unambiguously between the various categories of active nuclei on the basis of profile shape for the NLR. It seems that either kind of Seyfert galaxy usually has FWHM > 250 km s^{-1}, while starburst NLR are less than this (Feldman *et al.* 1982), but this rule is not inviolate (Phillips, Charles & Baldwin 1983).

Because the NLR is a much larger region than the BLR and has a narrower profile that is easier to center upon, the narrow-line profile is normally taken to define the systemic velocity of the quasar. Relative to this velocity, a number of characteristic asymmetries are observed. When asymmetric, both broad- and narrow-line components tend to extend more to the blue of the systemic velocity than to the red (Whittle 1985). It cannot be assumed that the observer has a preferential location, so interpretation of these asymmetries must produce a model such that an observer in any direction would see a similar result. As detailed more in Chapter 9, the models resulting require systematic outflow of gas from both BLR and NLR.

8.8   Other emission lines

The discussion so far has emphasized the emission lines which are most conspicuous and, therefore, most readily observed. As instrumentation improves, many other lines become observable even if weak. Understanding the ionization and excitation mechanism leading to any feature teaches us a little more about the conditions within quasars. The features whose interpretation is the most complex are those from the various ions of iron. Fe lines have been observed from Fe II, certainly to Fe x, and probably to Fe xIV. The higher ionization lines are useful in probing the conditions within the regions closest to the ionizing source. When done, this also confirms the picture of gas outflow, with lines of high ionization having greater widths and blueshifts relative to the systemic velocity than lower ionization lines (De Robertis & Osterbrock 1984).

At the other extreme of ionization, an entire array of Fe II lines blends and confuses many parts of the spectrum. Understanding these lines is necessary, if only to correct for their presence in order to locate the real continuum (Wills, Netzer & Wills 1985). Because these lines can be excited by resonance fluorescence, their presence may indicate the presence of high density cloudlets. An intriguing result that indicates the utility of wide ranging spectroscopic investigations is the tendency for Fe II lines to be much stronger in Seyfert 1 nuclei which are weak radio sources compared with those which are strong radio sources (Osterbrock 1977). That details of the optical spectrum relate to characteristics of the radio properties is amazing enough even though there is no good understanding of what causes this correlation (Steiner 1981).

Helium lines are often sufficiently detectable to be usable in refining models of the emission regions. Being produced by recombination processes comparable with those producing hydrogen emission, these lines would be expected to scale in intensity as the helium abundance scales to hydrogen, so the brightest helium lines are only a few per cent as intense as the Balmer lines. Interpreting the helium spectrum is complicated by various excitation effects that are possible, as well as by the unknown helium abundance (MacAlpine 1981, Grandi 1983). Using helium features requires some sophisticated interpretations; there are no really simple uses of these lines.

Because there are at least 26 elements in quasars, there are many more conceivable spectral features than commented upon here. The last ones worth mentioning for their relatively straightforward utility are the lines of very low ionization that are often conspicuous, including [O I] $\lambda 6300$, [N II] $\lambda\lambda 6548,6562$, and [S II] $\lambda\lambda 6717,6731$. When these lines are strong along with the higher ionization features, as is often the case in quasars, it is another indication that a wide mix of regions is being seen. Such features are useful for tuning ionization models, although there have been fundamental disagreements as to the source of ionization. While these features were once attributed to collisional processes associated with shock heating (Heckman 1980), the same data now seem better explained by radiative ionization (Halpern & Filippenko 1984). The strengths of these low ionization features provide a straightforward diagnostic of whether the ionizing continuum is a power law (presumably non-thermal) in contrast to ionization by hot stars. A power law continuum, having relatively more low energy photons, causes these features to be much stronger than in ordinary H II regions ionized by stars. This provides the basis for a spectroscopic classification of nuclei based on the nature of the ionizing continuum (Baldwin, Phillips & Terlevich 1981).

How far one should or can go in deducing detailed models of the physical conditions in the ionized regions of quasars is a troubling question. The hard work put in so far has shown how terribly complicated these regions are, that they differ from quasar to quasar, and that there seem to be no simple parameters that scale with quasar luminosity. Nothing analogous, for example, to the stellar Hertzsprung–Russell diagram is on the horizon. Analysis of emission line spectroscopy has played a vital, fundamental role in yielding order-of-magnitude descriptions of the conditions within quasars. Improving the quality of these descriptions may prove to be a chore that frustrates even the cleverest modellers. The basic problem is that so many things are going on at once, all of which contribute something to the net spectrum observed. Whether that spectrum can be disentangled completely is not at all obvious, but the unique role of these techniques in producing quantitative parameters for quasar interiors makes the task a vital one.

9

QUASAR STRUCTURE

9.1   Introduction

It should be clear from the discussions in the preceding chapters that an overwhelming amount of information is now available for describing quasar properties. Observationally, the study of quasars has been a great success. Also, it should be no surprise that, as the data have accumulated, it becomes more difficult to produce models that can explain everything. As might be expected for the most energetic objects in the universe, quasars are complex. This should not be a source of discouragement. It is not necessary to understand all details of the solar surface to know why the Sun shines. It is not necessary to understand all sedimentary rocks to know why continents drift. It is not necessary to memorize the taxonomy of all living creatures to realize why evolution occurs. When we are after the fundamental understanding of why something happens, all of the details are not required. In the study of quasars, we are still struggling to the point of knowing which details can be safely ignored, and it is for guidance in this regard that existing models are most useful.

The single most significant observational datum about quasars is that their spectra are so extraordinarily similar, even over ranges of 10^7 in luminosity, for objects separated by more than ten billion light years in the universe. Low redshift Seyfert 1 galaxies, as a group, have spectra virtually identical to those of quasars with $z > 2$, even though the latter are systematically more than 10^4 as luminous and are seen at look-back times exceeding 10^{10} years. Just within the local quasar luminosity function, nuclei with fundamentally similar spectroscopic properties are seen encompassing a range of 10^4 in luminosity. Differences in relative spectral line fluxes, line widths and equivalent widths are less than factors of two. Comparing emission line luminosities and the optical non-stellar continuum luminosities with X-ray luminosities shows that these scale together in luminosity (Shuder 1981, Yee

1980, Kriss & Canizares 1982). Lest it be suspected that these correlations somehow are a selection effect, because quasars are chosen spectroscopically by having strong emission lines and non-thermal continua, samples found in completely different ways fit right in. One of the more important uses of X-ray surveys was to show that quasars found from their X-ray properties look just like those discovered by their optical properties, even though the survey wavelengths differ by a factor of 1000.

As discussed previously, this continuity of properties argues in favor of cosmological redshifts for quasars, since the astrophysical mysteries posed by the Seyfert 1 galaxies are just as serious as those of the high redshift quasars. Their spectral similarities imply that they have fundamentally similar central engines, but ones whose power throttles can be turned up by a factor of 10^7 without changing the kind of physical processes taking place. The distinction of all quasars is their ability to generate extraordinary luminosities within small volumes. Luminosity generation can be accounted for in many ways; in principle, simply putting enough stars together could create arbitrarily high luminosities. The most shocking discovery for quasars came with the realization that luminosities observed were arising from volumes far too small to contain the stars needed to account for energy generation by nuclear processes. That is why quasars were recognized as being really different from processes normally associated with galaxies; something other than the energy of nuclear fusion had to be used.

The simplest way now known to produce such an engine is to utilize the gravitational energy released by accretion of matter onto a massive object, an object usually taken to be an MBH. Quasars are known to occur in the centers of galaxies, locations that are well defined, symmetric centers of gravity. There are various plausible scenarios by which massive, condensed collections of matter should develop in galactic centers (Rees 1984, Shapiro & Teukolsky 1985). Hypothesizing that this matter eventually reaches equilibrium in the form of a single MBH is desirable because the properties of such an MBH are well defined. Model making can proceed in search of predictions that can be confirmed or denied observationally. At the present stage of quasar astronomy, the search continues for a proof that this model is fundamentally correct.

9.2   Fundamentals of accretion models

For accretion processes, the basic energetics are simple, but extreme complications arise in describing how the kinetic energy gained by accreting material is transformed to the radiation that is eventually observed. The fundamentals start by considering a mass m falling to radius R toward another mass M. If the fall is from infinity, m reaches R with a velocity v that is the same as the escape velocity from M at distance R, or $v = (2GM/R)^{0.5}$ within newtonian gravity. It can be seen from this what a powerful source of energy accretion can be. Consider neutron stars, objects whose existence is not questioned, of $M \approx 4 \times 10^{33}$ g and $R \approx 10^{6}$ cm. The kinetic energy contained by mass m reaching the surface of this neutron star is $mv^2/2 = 2.7 \times 10^{20}m$ erg. This kinetic energy is equivalent to $0.3\ mc^2$, and by eventual thermalization of the high velocity gas, or by various non-thermal processes, this kinetic energy can convert to radiation. A mass m can thereby yield 30% of its total annihilation energy just by falling.

Black holes (of the same M) have no efficiency advantage over neutron stars as long as the conversion of kinetic to radiative energy occurs at no smaller R. This is an important point for those with reservations about the existence of black holes. They are not necessary for accretion to take place as a powerful energy source. The conceptual advantage of black holes is that single ones can be of unlimited mass; achieving large concentrated masses with neutron stars would require clusters of such stars in contact but not coalescing. The 'Schwarzschild radius' of a black hole is defined as the point at which escape velocity equals light speed, or $R_s = 2GM/c^2$. In principle, M and R_s are not limited in size. All observable processes that will be discussed take place far outside the Schwarzschild radius. It does not make much difference for the models, therefore, exactly how the structure of the accretor is visualized. Obviously, the major goal of efforts to understand quasars is to determine what this accretor really is and how it came to exist. Unfortunately, observations and models so far available have to start with an assumed accretor, and this is usually considered as a single MBH.

Most observations that attempt to determine the structure of a quasar discuss one or all of four different zones. These are: (1) the accretor itself; (2) the region surrounding it where kinetic energy of incoming matter is transformed into radiation or outward moving matter winds and beams; (3) the region where broad permitted emission lines arise, the BLR; (4) the lower density region where forbidden lines are seen and mass motions are slower, the NLR. The accretor is the primary, smallest zone. It is assumed that it is centralized and localized, so that the remaining zones are distributed around it. That is, no complicated models exist in which there are multiple accretors mixed in with the other zones to be described.

9.3   Limits on the properties of a massive black hole

The first objective should be describing the accretors. Are they similar in all quasars? Are some quasars brighter than others because the accretor is more massive, or because the rate of accretion is greater? Do all galaxies have dormant or dead quasars in their centers? Are there quasars that are not associated with galactic nuclei? These questions couple with other major mysteries: why were quasars different in the past than now; what are the major forms of matter in the universe; when and where were the first generations of stars in the universe? None of these questions can be satisfactorily answered without knowledge of how much matter is contained within quasars.

It is never expected that the accretor can be detected by anything other than the gravitational effects it produces, so diagnostics have to be attempted within other zones to measure these effects. The smallest size scale that could be measured from any radiation in the quasar is that zone corresponding to the accreting matter that radiates closest to the event horizon. How small this scale is depends not only on the size of the accretor, but on how close to the Schwarzschild radius of a black hole one could expect radiating material to be located. The canonical value is a few Schwarzschild radii. A lot of factors affect this number. Material too close to the hole cannot be in stable orbit long enough to transform kinetic to radiation energy; relativistic time slowing effects and gravitational redshifts affect matter close to a Schwarzschild radius; there are such a variety of forms of radiation that all would not arise literally at the same radius.

In many quasars, the continuum has been observed to vary in flux at all parts of the spectrum on time scales of less than a year. Variability arguments relate to size in a qualitative, order-of-magnitude way. If a variable continuum is observed, the varying portion has to be significant compared with any stable portion, or else the variability would not be noticed. Some kind of signal, or process, or change in conditions in the quasar has to stimulate the variability. Such a change cannot propagate faster than the speed of light, so a qualitative statement can be made that the scale size over which the variable continuum arises cannot be larger than the distance light can travel in the time required for the variability to occur. It was once hoped that variability in the observed continuum luminosity would reveal a scale size for the region generating the luminosity consistent with the immediate vicinity of an MBH. This test requires luminosity that is isotropic, so that the observed measure indeed refers to the luminosity that predominates. As a result, any variability in non-thermal luminosity is suspect for this test because such luminosity can be enhanced by relativistic beaming. The observed output would have time scales changed and would not be representative of the true source luminosity. Varying non-thermal luminosity can represent flares taking place in a dense region far from the location of ultimate energy generation. For these reasons, use of variability arguments to constrain the size of the nuclear engine requires the observation of a variable continuum produced by thermal processes. Because it is natural to suspect that the hottest gas is closest to the MBH, being able to test the observed X-ray continuum for variability was an eagerly awaited objective. By observing hard X-rays (>2 keV), there would be confidence that the observed continuum emerged from the nucleus without being affected by absorption. As a result, variations seen would be caused by variations in the source of primary photons.

When these observations were made (Tennant & Mushotzky 1983), it was found that most quasars are not characterized by rapid X-ray variability. This was disappointing in context of qualitative requirements of MBH models for the size of the radiating region closest to the MBH. Taking a radius of 10 GM/c^2 for the smallest volume which should produce observable radiation, it is seen that this size is remarkably small even for large MBH. For $M = 10^{10} M_\odot$, this emitting region has radius 1.5×10^{16} cm, or 5×10^5 light seconds. To look for evidence of such a volume, variability on a scale less than 10^6 s is required. Such variability was not observed in general for the X-ray continua of quasars. The most rapid variability seen did have 100 s time scale, but that was observed only in two very low luminosity quasars. These objects, the Seyfert 1 nuclei of NGC 4051 and NGC 6814, have bolometric luminosities of about 10^{43} erg s^{-1}. For MBH models radiating at the Eddington limit, holes of only $10^5 M_\odot$ are needed, for which the size scale is 5 light seconds. The holes could be there and accreting at very low efficiency, however. In fact, the results of quasar evolution, deduced in Chapter 6, suggest that these nuclei were once 100 times as luminous, so there should be an MBH of $10^7 M_\odot$ there now. The radiating volume about such a hole would have radius 500 light seconds, so the observation of 100 s variability in these cases is consistent with expectations of MBH models.

The Eddington limit and observed luminosity constrain the minimum mass of any MBH in a quasar. The maximum mass can be constrained by the dynamics of the BLR. It is not known what is the source of kinetic energy for the clouds of the BLR. Arguments summarized in Section 9.5 indicate that these clouds are accelerated outward, so that their velocities represent ejection by radiation pressure. In that case, they must be moving at least at the escape velocity from the MBH. Escape velocity is given by $(2GM/r)^{0.5}$, with r the radius from which the clouds must escape. This r is the scale size which is set by the radius of the BLR, a scale size determinable from the ionization parameter and BLR variability. For one quasar, the nucleus of NGC 4151, both approaches give the consistent result that r is about 10 light days, or 3×10^{16} cm, for the clouds emitting C IV. Maximum outflowing velocities of the observed clouds, from the total width of the emission line profile, are 10^4 km s^{-1}. Setting this equal to the escape velocity gives $M = 10^7$

$M_\odot$. This is an upper limit because the observed cloud velocities may greatly exceed the escape velocity.

Alternatively, it might be contended that the cloud velocities represent a dynamical equilibrium such that the clouds are not escaping the nucleus, but have a velocity dispersion set up by the gravity of the hole. In that case, the cloud velocities can be used with the BLR radius to determine the dynamical mass of the hole. Using the virial theorem or assuming keplerian rotational velocities leads to relations between hole mass and cloud velocity within a factor of two of that given previously for escape velocity. In this case, however, the relevant velocity is the velocity dispersion or most probable velocity shown by the profile. This is closer to 10^3 km s^{-1} than 10^4 km s^{-1}. Once again, an upper limit to the hole mass is found, because of the likely possibility that all of the velocity dispersion is not gravitationally induced. This upper limit is close to $10^6 \, M_\odot$ for NGC 4151.

Observations of the BLR in NGC 4151 set, therefore, an upper limit to the MBH for this object of $10^7 \, M_\odot$. The lower limit, from the Eddington limit applied to a luminosity of 10^{44} erg s^{-1}, is $10^6 \, M_\odot$. The observed variability time scale of NGC 4151 in the X-ray continuum is one light day, yielding a minimum scale size for the X-ray emitting region that is a factor of 200 larger than that associated with a $10^7 \, M_\odot$ black hole. In sum, the X-ray continuum in NGC 4151 arises from a region a factor of 10 smaller than the BLR but at least a factor of 10 larger than the expected emitting radius about the MBH in that nucleus. (It is interesting that the upper limit to the MBH size is less than that which would be deduced by the conclusions of pure luminosity evolution for quasars. In the scenario whereby the local quasars seen today have always been quasars and were once 100 times brighter, NGC 4151 has to have a hole of $10^8 \, M_\odot$ to account for its past luminosity. As long as we deal in orders of magnitude, this kind of consistency check is not particularly meaningful. It does illustrate how various interpretations have to be examined in context of other conclusions.)

The kind of analyses reviewed are not available for any very luminous quasars, so we cannot check on the objects which should have much larger MBH than these Seyfert nuclei. More fundamentally, the indirect arguments given are only consistency checks on the MBH models; they do not prove the existence of an MBH, but only constrain its size within the hypothesis that it is present.

9.4   The accretion volume

While the hypothesis of an MBH provides a simple explanation of the ultimate power source of a quasar, exceptional complications are found when pondering how the gravitational energy released by accretion is transformed into radiation or relativistic particles. These crucial transformations occur in dense material as close as a few Schwarzschild radii to the MBH. Difficulties in interpreting observations arise because the accretion models are not unique; many different configurations of accreting material can be postulated, with various temperature distributions or non-thermal generators in each. The form of the accretion volume depends on the rate of accretion compared to that allowed by the Eddington limit, and on whether the accreted material has significant angular momentum.

Models which have been most thoroughly explored are based on accretion of matter with angular momentum that leads to the formation of an accretion disk (Begelman 1985). The remaining crucial parameter is whether the MBH is accreting at a rate much less than corresponds to the Eddington limit; if so, the 'thin disk' models apply. For interpreting a spectrum, even thin disk models can be legitimately modified to yield a wide range of consequences. The thermal structure is complex, hotter close to the MBH, so the observer sees the combined luminosity from regions of different temperature. This could mimic a power law spectrum, even if totally thermal in origin. Gas should boil off from the hot disk, leading to a corona around the thin disk, also with its own complicated thermal structure. Depending on the magnetic field structure, non-thermal mechanisms can be assigned more or less importance. Furthermore, any thin disk model is subject to orientation effects. Not surprisingly, the disk would appear differently if viewed face-on compared with edgewise.

The disk models with the simplest observational consequences are the 'thick disk' models, for which accretion rates are high compared to what the hole can swallow. In this case, radiation pressure inflates the disk, it is optically thick, and there is the hope that the observer might see it as a large object with a single temperature. The temperature would be that of the outermost optically thick layer, in exact analogy to why a star can be characterized by a single surface temperature. In fact, thick disks in the limit approach a configuration much like a huge star; very hot at the center, but with temperatures $\sim 10^4$ K on the outside. These are the kinds of disks that some observers think explain the 'blue bump' in the quasar continuous spectrum.

Analogous accretion zones can be imagined with spherical accretion, in which the accretion is three dimensional, and a disk structure never develops. Depending also on the accretion rate, spherical accretion could lead to an extreme, optically thick configuration resembling a thick disk. Alternatively, optically thin spherical accretion could allow the hot gas at the center of the accretion volume to be seen, explaining the power law X-ray continuum as thermal emission from different temperatures (Meszaros 1983). In spherical accretion scenarios, as in disk scenarios, there is virtually unlimited room for adding non-thermal modifications.

In sum, the problem with accretion models is not in coming up with an explanation of what is observed. The problem is that there are too many ways to explain what is observed; no models are as yet unique. Yet, we very much desire a quantitative model such that believable size constraints can be placed on the MBH itself. A meaningful consensus of the wide variety of accretion models is that substantial thermal continuum should be generated in the accretion zone. Whether the escaping photons are reprocessed by interaction with relativistic particles to produce non-thermal radiation is a complicating unknown. Even then, there is the hope that this reprocessing might take place close to the accretion region. For the observer, the conclusion is that the best way of probing close to the accretion volume is to study the continuous spectrum. Defining the place where this is produced must certainly give useful constraints, at least, on the size of the accretion disk or sphere.

As already described, variability arguments and ionization modelling can give measures of the size of the BLR surrounding the continuum source. By definition, this thereby constrains the size of the continuum source to something smaller than the BLR. If one is confident the results are not much contaminated by relativistic beaming, source variability in the continuum also constrains the size of the continuum source. These constraints are not yet very meaningful, however, as short time scale continuum variability is usually associated with blazar sources that are suspect of beaming. Making much headway in locating confirmable examples of rapidly varying thermal sources will require careful monitoring of X-ray spectra – a task for AXAF.

There is one object for which the size of the continuum-generating region has been directly measured. This is in the very nearby quasar which resides in the nucleus of the galaxy M81. Although very low in luminosity, the nucleus of M81 counts as a real quasar. It possesses all of the required symptoms, including a BLR and X-ray luminosity scaling by the expected amount (Peimbert & Torres-Peimbert 1981, Elvis & Van Speybroeck 1982). I consider the nucleus of M81 to be the closest quasar. There have been several suggestions of quasar-like activity in the nucleus of our own Milky Way Galaxy (Serabyn & Lacy 1985), based on the deduction that a bolometric luminosity of 10^{40} erg s^{-1} and mass of 10^6 $M_\odot$ are within the central 3 pc. Neither a central X-ray source nor a localized BLR is seen in the nucleus; activity is noted from radio continuum and infrared emission line observations with very high spatial resolution. There is no way to calibrate what is seen in the nucleus of the Milky Way, because comparably low level activity cannot be observed anywhere else.

The low luminosity limit for objects with observable BLR in the local quasar luminosity function of Figure 5.2 is comparable with the nucleus of M81, or about 10^{40} erg s^{-1}. Although similar to the nuclear luminosity quoted for the Milky Way, the activity in M81 arises in the very small volumes which characterize quasars. The compact, flat spectrum radio source in the center of M81 has been mapped with VLBI (Bartel *et al.* 1982). It has a diameter of about 0.01 pc, such a small value being resolvable because the galaxy is only 3 Mpc distant. The source is oblong in shape and is a different size at different frequencies, as expected for optically thick synchrotron emission. Quite a lot of important conclusions can be drawn from this one VLBI observation of M81. Because the continuum region is non-thermal, it should be outside the accretion disk which generates the primary thermal photons and relativistic particles. Even though it is the radio continuum that is seen, we can expect that the optical and X-ray continua are produced within a volume no larger than that mapped. The radius scale of 0.005 pc, or 5 light days, is close to the size of the continuum region that must exist in NGC 4151, for which independent studies yielded a BLR of radius about 10 light days about the continuum source. It is not possible to say that the M81 observation shows an actual accretion disk about an MBH, but it provides the most quantitative limit yet available for the size of one. We do not know yet how this radio source in M81 is placed with respect to the BLR and the source of X-rays. Perhaps it represents the volume where all of the non-thermal luminosity arises, and so is really the ionizing source for the BLR. Perhaps, at the other extreme, it is a plasma cloud that has already been ejected from the vicinity of the central engine, so is outside of or larger than the BLR. Those questions remain; nevertheless, we can now honestly say that the inside of a quasar has been imaged, and that is major progress.

The search for any proof that the accretion models are correct in their basic assumption of an MBH is still on. This explains why it is so important to find evidences in quasars of oriented energy release. If the energy release is attributed to the consequences of a single object, there is hope that this object might define a reference frame. In the case of the MBH, this reference frame would be established by the angular momentum vector of the accretion disk. Evidence that such a coherent frame is established could then be taken as evidence for the MBH plus accretion disk model. That a preferential direction exists for at least a few quasars is established by the well collimated particle beams that emerge from some nuclei. Generically called jets, these are directed winds of plasma and magnetic fields, possibly moving at relativistic velocities, which are most often seen via their continuum radio emission (Bridle & Perley 1984, Phinney 1985). Observed jets range from pc to Mpc lengths. Understanding the propagation and collimation of these jets is the basic challenge of explaining extended radio sources (Begelman, Blandford & Rees 1984). The conventional model has collimation being provided by the walls of the deep funnel present in a thick accretion disk; explosive release of matter from the vicinity of the MBH is routed out the funnel. There are many steps left before a fully consistent model exists. A major question is how and why a few radio bright quasars are able to put most of their energy into the jet rather than into the thermal and non-thermal radiation normally seen.

An important caution is that these jets are seen in less than 0.1% of known quasars. It may be that uniformitarianism is being pushed too far to require that jet production has any relation to mechanisms in the vast majority of quasars without jets. For the individual cases where jets are there, studying them is vitally important to understand the mechanisms for accelerating and collimating particle beams.

It is not appropriate, however, to assume that orienting or beaming mechanisms are in all quasars unless other evidence can be found for preferential directions defined in these quasars. Various searches are underway for such evidence. These include efforts to detect polarization. Even if the mechanism for polarization is not understood, the existence of polarization means that something has an alignment. The internal structure of the quasar is not random and isotropic. Polarization caused by electron scattering or by dust scattering would be evidence that the electron distribution or the dust has axis of symmetry and an alignment with respect to the interacting photons. Any correlation between the polarization vector and other non-isotropic properties, such as the direction of radio structure, confirms the evidence for alignment (Schmidt & Miller 1985).

Except for the small minority of cases where jets, other radio structure or polarization gives evidence of a preferential orientation, there is no proof that the core of a quasar must contain a single accretion disk. The evidence of the continuum and emission lines is that there is a luminous continuum source which is small, and that this source is surrounded by emitting clouds whose density and velocity decrease with increasing distance from the ionizing source. Although necessary for organization, presenting this discussion in terms of discrete, well defined zones in a quasar is unrealistic. There are many reasons to believe that these zones merge and blend in complex ways. In a perverse fashion, such effects can make many models 'right' for at least some quasars. Matter boiling off and condensing in the vicinity of the accretion disk could form BLR clouds, which could be immersed in ionizing radiation from nearby non-thermal sources. This is far from the simple picture of a BLR shell surrounding a point-like ionizing source. Mixed in and passing through this blend would be the relativistic clouds carrying beamed non-thermal radiation.

The difficulty in disentangling all of this to first order is insufficient spatial resolution. Even the best available VLBI techniques only yield resolution limits above 1 pc on distant quasars. At this scale, it is not possible to determine if the non-thermal source is inside, mixed with, or outside the BLR. Some improvements will come with VLBA. The superluminal sources are particularly promising, usually showing a core and a jet on pc scales. In preparation for improved VLBA maps I encourage line variability monitoring and careful ionization modelling of the BLR for those superluminal sources with conspicuous BLR, particularly 3C 120, 273, 279 and 345.

9.5   The broad-line region

As emphasized many times in this volume, the presence of broad permitted emission lines is the best criterion by which a quasar can be recognized. Whether quasars exist with obscured BLR, or whether accreting MBH can be found without producing BLR, are issues to investigate in context of Seyfert 2 galaxies and radio bright quasars. At any rate, we know that most currently cataloged quasars have measurable BLR. Beyond providing a recognition criterion, the great importance of the BLR is that, so far, it is the closest region to the quasar's central engine that can be quantitatively modelled to yield its structure, size and kinematics. Understanding the BLR is a necessary route to understanding the next regions interior to it, where the continuum arises.

Even though the primary continuum radiation is supposedly generated closer to the accretor than the BLR, there is a well demonstrated correlation between the presence of that continuum and the existence of the BLR. Obviously, ultraviolet ionizing radiation is required to produce a BLR. So the latter could not exist without the former. Yet, except for the handful of blazars, strong continua are not seen without accompanying BLR. The correlation was found to go further than that once the X-ray data came available. Hard X-ray radiation should be as fundamental an observable radiation as there is, and there is no *a priori* reason why this has to bear any relation to the ionizing continuum. Yet the correlation of HX and luminosity of the BLR is well established (Lawrence & Elvis 1982, Kriss & Canizares 1982). In fact, this correlation led to the discovery of galactic nuclei with weak BLR that had been overlooked by optical astronomers. A class of strong X-ray galaxy was found, confusingly named the NELG (narrow emission line galaxies). Upon careful observation, these were found to have broad but very weak Hα lines after all (Veron *et al.* 1980). This discovery proved that the correlation between strong HX and the BLR actually had predictive power. It is not appropriate, therefore, to consider the region of continuum production and the BLR as independent zones without some basic dependence upon one another. The simplest picture relating them is to imagine that the continuum originates in optically thick material close to the accretor, ideally in an accretion disk to simplify the models, and that material boils off this disk to produce the BLR. One would never be found without the other in this scenario.

Combining X-ray, optical and ultraviolet observations with careful ionization modelling can lead to major progress in understanding the BLR. The most thoroughly analyzed quasar is the nucleus of NGC 4151. Results of Holt *et al.* (1980) and Ferland & Mushotzky (1982) yield a picture of the BLR structure relative to the continuum source. The BLR is a 'leaky absorber' made of small but dense clouds, each of which is smaller in projected size than the continuum source. These clouds produce the emission lines and move at high velocities. The ensemble of clouds covers about 90% of the continuum source at any time (the 'covering factor'), but variations arise as individual clouds move about. The continuum can be seen only through the holes between the clouds. A source of major satisfaction is the consistency between this description of the BLR and that derived by careful comparison of continuum and emission line variability seen by the IUE (Ulrich *et al.* 1984). Both studies yield a scale size of about 10 light days for the separation between continuum source and the BLR where C IV emission arises.

In comparison with other quasars, this picture for NGC 4151 differs primarily in that the BLR is closer to the continuum source (higher ionization parameter) and covers more of the source than in higher luminosity quasars. Studies of both X-ray absorption (Lawrence & Elvis 1982, Reichert *et al.* 1985) and ionization parameter (Mushotzky & Ferland 1984) indicate that the BLR enlarges and covering factor decreases as quasar luminosities increase.

Properties of the individual clouds in the BLR are such that these clouds are optically thick via free–free absorption to radio frequency photons. This means that any compact radio source (scale of order 1 pc) which is seen must either be visible through holes between the clouds or must arise outside the BLR. If it arises outside of the clouds, it arises, by definition, in the NLR, and it is necessary to explain how relativistic particles are energized in that region. It seems clear that the radio power on much larger scales (100 pc to kpc) does correlate with the luminosity and kinetic energy of the NLR (Meurs & Wilson 1984). Even with this correlation, there are interpretations in opposite senses. One is that the kinetic energy already in the NLR stirs up sufficient shocks and magnetic fields to radiate non-thermally. The alternative is that the same particle beams carrying energy from the core to the NLR produce the radio luminosity and deposit the kinetic energy into the NLR. The latter interpretation is favored by observations that show coherent radio structures in the NLR (Wilson & Ulvestad 1982). With this evidence, it can be argued that the shape and thickness of the BLR plays a major role in whether or not a quasar is visible as a strong radio source. Correct modelling of the BLR to include explanation of quasar radio properties will probably show that some BLR also have a disk-like configuration (Osterbrock 1984).

Emission line profile data for the broad lines provides the only kinematic data from inside the quasar. These profiles show how gas clouds inside the BLR are moving. It is not known yet why they move as they do, but the potential is there for a lot of dynamical understanding. Extensive profile data have been collected and tested for correlations that would be expected with various nuclear models (Shuder 1982). Quite a few observations indicate that the motions in the BLR have a substantial component of radial motion away from the center. The observations definitely show that line widths increase for lines of higher ionization and higher density (De Robertis & Osterbrock 1984). In a simple model with the ionizing source at the center of the BLR, these results mean that the cloud velocities are faster closer to the center. That these motions represent systematic outflow is an interpretation dependent upon the observation of asymmetries in the broad-line profiles.

These line asymmetries are in the opposite sense in different features, and a self-consistent interpretation requires careful consideration of the dust absorption and radiative transfer effects described in Chapter 8. The Ly α line often is asymmetric to the red, in that the blue side of the profile seems cut or absorbed away. That can be explained by realizing the high optical depth to Ly α of individual BLR clouds; Ly α leaves clouds preferentially on the side of the cloud closest to the ionizing source. Those clouds on the far side of the continuum source, the receding clouds, are preferentially seen. In a contrary sense, the Hα and Hβ features are able to pass through single clouds and are more affected by absorption, in dust or gas, over the total path length through the BLR. These photons are preferentially seen from the near side, approaching clouds and, in fact, do show preferential asymmetries to the blue.

Other observations exist of the motions within the BLR. These are observations of absorption lines, which can only be seen in projection against the continuum source. There is a category of quasars that has very deep and broad absorption lines from highly ionized ions: C IV, S IV, N V, O VI, which extend over tens of thousands of km s^{-1}. These absorption lines are so deep that they sometimes obscure the entire continuum and most of the broad emission lines from the same ions. A few per cent of quasars with redshifts adequate to see such lines if they are present actually show these features. These quasars are called 'broad absorption line', or BAL, quasars (Turnshek 1984). Because the absorption cuts out broad emission lines as well, the absorbing clouds are occulting not only the continuum, but also much of the ionized gas. It has not been possible to determine just where these BAL clouds are located relative to the continuum source and BLR, other than that they obviously must be on the near side. The point of discussing the BAL quasars here is that the BAL are always blueshifted relative to the center of the broad emission lines, usually beginning just at the central redshift and extending up to 60 000 km s^{-1} to the blue. This proves that incredible gas velocities are present, and all are coming toward us relative to the continuum source. This is also evidence that high velocity gas is flowing away from the continuum source. Taken with the evidence given previously that indicates gas in the BLR closer to the continuum has the highest velocities, we would conclude that the bluest part of the BAL are closest to the continuum source. So far, the BAL quasars have only been seen at high redshifts and high luminosities.

Given the evidence that there are systematic radial motions of outflow in the BLR, attempts to use the broad-line profiles to measure gravitational potentials in quasars probably are not valid, except to set crude upper limits. As it is possible that most of the observed line widths reflect radiative acceleration of the clouds, and that the clouds may well be unbound and flowing out of the quasars, the observed widths give upper limits to the gravitational potential that could be far too large. Although line width data is still the only way to estimate the mass really present in the quasar nucleus, these measurements critically depend on scale size for the BLR radius. Much more data are needed from variability monitoring and ionization modelling to see how results such as those for NGC 4151 change with quasar luminosity. The preliminary indications that the BLR enlarges with luminosity are in the correct sense; the MBH mass could increase without changing the gravitational effect on BLR velocities.

Existing correlations show that high luminosity decreases the covering factor, as if the more powerful continuum were able to drive out the clouds of the BLR. Some interesting geometrical deductions arise from this; if the volume of the continuum zone and that of the BLR simply scaled together, but the number and size of individual clouds did not increase, then the covering factor decreases with luminosity. Even if the continuum zone remained the same size, but the BLR gets larger, the covering factor also decreases. Only if the continuum source enlarged but the BLR remained the same would the covering factor remain the same. The correlations are best explained, therefore, by an enlarging BLR with number of clouds per unit volume diminishing with increasing luminosity. This implies that expansion induced by radiation pressure is an important scaling with luminosity, which predicts that line variability will take longer in more luminous quasars.

The BLR must be a very complicated place. Lines of widely different ionization potentials are seen, and the degree of ionization depends not only on distance from the continuum source, but also upon where within an individual cloud a line arises. The side of a cloud near the ionizing source shields the interior of that cloud, so that low ionization lines like Fe II could arise deep within the clouds of the BLR. This means that some low ionization features could arise closer to the source than high ionization features. Dust effects probably are important within clouds of the BLR, but the difficulty of disentangling dust and high density effects makes this uncertain. Whatever dust effects occur in the BLR occur within individual clouds and not between clouds. The inter-cloud environment must be very hot in order to keep the clouds confined (Weymann *et al.* 1982). Such an environment would not interfere with the passage of optical photons and would not allow dust to survive.

9.6   The narrow-line region

The last zone recognizable as an unusual region associated with most quasars is the NLR. Herein, the mass of gas is much greater than in the BLR, but individual clouds have lower velocities and spread over a far larger volume. The NLR is so large that it can be resolved in many Seyfert galaxies, and a canonical radius is a few hundred pc. Because the volume is so much greater than the BLR, several orders of magnitude more gas is contained in the NLR. The simple equations of emission line spectroscopy can be applied to the NLR, because densities are low enough that collisional excitation and radiative transfer effects do not distort Balmer line ratios. The NLR is best seen in the light of [O III] $\lambda\lambda 5007,4959$, this being the strongest forbidden line. There are no strong forbidden lines from the NLR in the ultraviolet spectra of quasars, so we do not really know the nature of the NLR in the high redshift quasars.

As has been the case throughout this discussion, the ultimate objective is to understand the central engine. In this regard, the main reason for studying the NLR is as a diagnostic of what escapes from the BLR to be seen in the NLR. Keeping track of the quasars' total energy budget requires us to consider everything. For nearly all quasars, the NLR is the end of the line; that is, the last volume containing any evidence of unusual events. The only quasars which are able to affect or release matter beyond the NLR are the few radio bright quasars surrounded by extended radio lobes.

It is the contrast between the appearance of the NLR and the BLR that has produced many of the details of the optical categorization of quasars. In objects with sufficiently low redshift for the necessary forbidden lines to be seen, virtually all quasars have an NLR. Relative to the luminosity of [O III], the BLR lines may be weak or strong. It is this apparent ratio of Balmer to forbidden lines that controls Seyfert classifications, such as Seyfert 1, 1.5, 1.8, etc. The Seyfert 2 galaxies have no detectable broad lines; obviously, they would not be recognized as distinctive at all if they did not have an NLR that was more luminous and kinematically active than 'normal' galaxies. In my opinion, the primary motivation for understanding the NLR is to search for quasars whose BLR is obscured. That is, can we find evidence of a category of quasars that has been overlooked because the characteristic signature of a BLR is missing? It is in seeking the answer to this that the relation between the NLR of known quasars and that of Seyfert 2 galaxies (and, equivalently, 'narrow-line radio galaxies') is important.

All quasars with BLR are clearly similar, and by studying the gradations from weak to strong BLR, compared with NLR, things can be deduced. Seyfert 2 galaxies, however, may not be related at all to the objects so far discussed as quasars, because Seyfert 2 galaxies do not show evidence of the intimately connected continuum and broad-line zones that have been described. We need to know how far we can go in simplifying the issue of quasars. Can one basic theory really explain them all? Can all quasars be accounted for by the effects of massive accretors in the nuclei of galaxies? Can any other process energize the centers of galaxies? It is to answer these questions that a variety of active galactic nuclei are considered, many of which have very different properties from others. Can these be related to the conventional quasar scenario, or not? Starburst galaxies were discussed in Chapter 7; the other most common category of active galactic nuclei is the Seyfert 2 galaxies. I remain very puzzled by these. Basically, the question is whether there is an ionizing continuum and BLR similar to that of quasars in the centers of Seyfert 2 nuclei, or whether the energizing of the observable NLR has been caused by some other process (Terlevich & Melnick 1985).

Arguments in favor of obscuration are based on extrapolations of effects observed in nuclei with BLR. The decreasing visibility of the broad lines is the basis of Osterbrock's spectroscopic classification to Seyfert 1.9. The relevant correlation is that the Hα/Hβ flux ratio from the BLR increases toward class 1.9, implying increasing dust obscuration (Osterbrock 1984). Correlations indicate that dust is likely to be associated with the NLR in quasars (Rudy 1984). It is a natural extension of this trend to imagine obscuration so severe that the BLR is blocked off, producing a Seyfert 2. Analogous reasoning is used by Lawrence & Elvis (1982), who note X-ray absorption as measured by the ratio of soft X-rays (0.5–3.5 keV) to hard X-rays (2–10 keV). The latter are not affected by absorption. As HX luminosity diminishes, absorption of SX increases. Among the most absorbed sources are the NELG, with very faint BLR in Hα and none in Hβ. Once again, a natural extension is that Seyfert 2, very weak in HX are heavily absorbed in SX, and that the same absorbing clouds obscure the BLR.

There are some contrary considerations. Seyfert 2 galaxies have even more luminous NLR than do the Seyfert 1, and these NLR are radiatively ionized (Koski 1978), so ionizing photons are certainly getting into the NLR. If the covering factor were unity, this could not happen. Also, some X-ray observers feel that the Seyfert 2 are so systematically underluminous in X-rays as to show no evidence of a quasar-like ionizing continuum (Kriss, Canizares & Ricker 1980). Finally, data exist indicating that Seyfert 2 have changed to Seyfert 1, and vice versa. There are several nuclei in which BLR have been observed at some times but not others. Taking the latter data literally gives a simple answer for the explanation of Seyfert 2; they are normal quasar nuclei in which the ionizing and X-ray continua are temporarily in a low state. They should not count in quasar statistics unless shining as quasars. When these continua brighten, the BLR is ionized and the appearance becomes that of a conventional quasar. Attributing data inconsistencies to variability is a convenient excuse, however, so such interpretations have to be taken with caution. Furthermore, two other separate indicators show systematic differences between the NLR of quasars and Seyfert 2 nuclei. The NLR lines are broader in Seyfert 2 (Heckman *et al.* 1981, Feldman *et al.* 1982), and the NLR are stronger radio sources in Seyfert 2 (Meurs & Wilson 1984).

NGC 1068 presents an excellent example of the confusion that can arise. This is a strong far-infrared source, with an infrared continuum extending beyond 100 μm. Initially, this continuum was attributed to the nucleus so had to be considered as part of the energy budget of whatever powers the nucleus. It now appears that the far-infrared continuum of NGC 1068 is not from the nucleus, but from an independent region of star formation in an extended disk surrounding the nucleus. The nuclear infrared continuum seems to peak at about 20 μm (Telesco *et al.* 1984). This nuclear continuum shows the silicate absorption features that are an unambiguous sign of dust, but the continuum arises from an unresolved zone ($<1''$) much smaller than the NLR. From these data, it appears that NGC 1068 indeed has a small central region with so much dust that a lot of continuum has been absorbed and is now re-radiating in the mid-infrared. A coherent interpretation would require, however, that this dust be restricted to individual clouds, with sufficient gaps between them for an ionizing continuum to escape. Confusion is compounded by the complex polarization behavior (Miller & Antonucci 1983).

If Seyfert 2 nuclei are considered to be obscured quasars, the net effect would be to increase the number of low luminosity, local quasars by about a factor of two. Very few analogs to Seyfert 2 have been observed among high redshift, high luminosity quasars. The obscuration, even if applicable at low luminosities, does not often affect high luminosity sources.

It will prove ironic if the Seyfert 2 galaxies are the category of active galactic nuclei which take the longest to understand. Because these were the first to be noticed; V. M. Slipher admitted in 1918 (Slipher 1918) that he could not account for the one he knew about, stating, 'The disk-form images of the chief nebular emissions . . . is peculiar to this nebula . . . Perhaps pressure increasing towards the center or nucleus of the nebula might be a sufficient cause for the great broadening of the lines . . .'. Slipher was a patient man, once spending 35 hours exposing a single spectrogram on NGC 1068. Perhaps we should extend such patience, and not expect all of the mysteries of quasars to succumb until a few more generations of astronomers have tried their luck.

REFERENCES

References with more than ten authors are listed as *et al.*

Chapter 2

Arp, H. C. 1981, *Astrophysical Journal*, **250**, 31.

Arp, H. C. 1983, *Astrophysical Journal*, **271**, 479.

Avni, Y., Soltan, A., Tananbaum, H. and Zamorani, G. 1980, *Astrophysical Journal*, **238**, 800.

Avni, Y. and Tananbaum, H. 1982, *Astrophysical Journal (Letters)*, **262**, L17.

Barbieri, C. and Romano, G. 1981, *Astronomy and Astrophysics*, **99**, 206.

Braccesi, A. 1983, in *Early Evolution of the Universe and its Present Structure*, eds. G. O. Abell and G. Chincarini, IAU Symposium 104, p. 23. (Dordrecht: Reidel).

Burbidge, G. 1979, *Nature*, **282**, 451.

Burbidge, E. M., Burbidge, G. R., Solomon, P. M. and Strittmatter, P. A. 1971, *Astrophysical Journal*, **170**, 233.

Burbidge, G. R., O'Dell, S. L. and Strittmatter, P. A. 1972. *Astrophysical Journal*, **175**, 601.

Burstein, D. and Heiles, C. 1978, *Astrophysical Journal*, **225**, 40.

Cecil, G. and Stockton, A. 1985, *Astrophysical Journal*, **288**, 201.

Clowes, R. G., Cooke, J. A. and Beard, S. M. 1984, *Monthly Notices Royal Astronomical Society*, **207**, 99.

Condon, J. J. 1984, *Astrophysical Journal*, **287**, 461.

Condon, J. J. and Mitchell, K. J. 1984, *Astronomical Journal*, **89**, 610.

Feigelson, E. D. and Nelson, P. I. 1985, *Astrophysical Journal*, **293**, 192.

Gioia, I. M., Maccacaro, T., Schild, R. E., Stocke, J. T., Liebert, J. W., Danziger, I. M., Kunth, D. and Lub, J. 1984, *Astrophysical Journal*, **283**, 495.

Greenstein, J. L. and Oke, J. B. 1970, *Publications Astronomical Society of the Pacific*, **82**, 898.

Griffiths, R. E. *et al.* 1983, *Astrophysical Journal*, **269**, 375.

Hawkins, M. R. S. 1983, *Monthly Notices Royal Astronomical Society*, **202**, 571.

Hayes, D. S. and Latham, D. W. 1975, *Astrophysical Journal*, **197**, 593.

Hazard, C., Morton, D. C., Terlevich, R. and McMahon, R. 1984, *Astrophysical Journal*, **282**, 33.

He, X.-T., Cannon, R. D., Peacock, J. A., Smith, M. G. and Oke, J. B. 1984, *Monthly Notices Royal Astronomical Society*, **211**, 443.

Heckman, T. M., Bothun, G. D., Balick, B. and Smith, E. P. 1984, *Astronomical Journal*, **89**, 958.

Hewett, P. C., Irwin, M. J., Bunclark, P., Bridgeland, M. T., Kibblewhite, E. J., He, X.-T. and Smith, M. G. 1985, *Monthly Notices Royal Astronomical Society*, **213**, 971.

Hoag, A. A. and Smith, M. G. 1977, *Astrophysical Journal*, **217**, 362.
Impey, C. D. and Brand, P. W. J. L. 1982, *Monthly Notices Royal Astronomical Society*, **201**, 849.
Kirshner, R. P., Oemler, A. Jr and Schechter, P. L. 1979, *Astronomical Journal*, **84**, 951.
Koo, D. C. 1984, in *Quasars and Gravitational Lenses*, Proceedings of 24th Liège Astrophysical Colloquium, p. 240.
Koo, D. C. and Kron, R. G. 1982, *Astronomy and Astrophysics*, **105**, 107.
Kriss, G. A. and Canizares, C. R. 1985, *Astrophysical Journal*, **297**, 177.
Maccacaro, T., Gioia, I. M. and Stocke, J. T. 1984, *Astrophysical Journal*, **283**, 486.
Margon, B., Downes, R. A. and Chanan, G. A. 1985, *Astrophysical Journal Supplement*, **59**, 23.
Neugebauer, G., Soifer, B. T., Miley, G., Young, E., Beichman, C. A., Clegg, P. E., Habing, H. J., Harris, S., Low, F. J. and Rowan-Robinson, M. 1984, *Astrophysical Journal (Letters)*, **278**, L83.
Osmer, P. S. 1982, *Astrophysical Journal*, **253**, 28.
Peebles, P. J. E. 1973, *Astrophysical Journal*, **185**, 413.
Peterson, B. A., Savage, A., Jauncey, D. L. and Wright, A. E. 1982, *Astrophysical Journal (Letters)*, **260**, L27.
Pica, A. J., Pollock, J. T., Smith, A. G., Leacock, R. J., Edwards, P. I. and Scott, R. L. 1980, *Astronomical Journal*, **85**, 1442.
Piccinotti, G., Mushotzky, R. F., Boldt, E. A., Holt, S. S., Marshall, F. E., Serlemitsos, P. J. and Shafer, R. A. 1982, *Astrophysical Journal*, **253**, 485.
Reichert, G. A., Mason, K. O., Thorstensen, J. R. and Bowyer, S. 1982, *Astrophysical Journal*, **260**, 437.
Savage, A., Trew, A. S., Chen, J. and Weston, T. 1984. *Monthly Notices Royal Astronomical Society*, **207**, 393.
Schmidt, M. and Green, R. F. 1983, *Astrophysical Journal*, **269**, 352.
Seldner, M. and Peebles, P. J. E. 1979, *Astrophysical Journal*, **227**, 30.
Shanks, T., Fong, R., Green, M. R., Clowes, R. G. and Savage, A. 1983, *Monthly Notices Royal Astronomical Society*, **203**, 181.
Smith, M. G. 1981, in *Investigating the Universe*, ed. F. D. Kahn, p. 151. (Dordrecht: Reidel).
Smith, M. G. 1984, in *Quasars and Gravitational Lenses*, Proceedings of the 24th Liège Astrophysical Colloquium, p. 4.
Soifer, B. T. *et al.* 1984, *Astrophysical Journal (Letters)*, **278**, L71.
Stocke, J. T., Liebert, J., Gioia, I. M., Griffiths, R. E., Maccacaro, T., Danziger, I. J., Kunth, D. and Lub, J. 1983, *Astrophysical Journal*, **273**, 458.
Sulentic, J. W. 1981, *Astrophysical Journal (Letters)*, **244**, L53.
Tucker, W. H. 1983, *Astrophysical Journal*, **271**, 531.
Tyson, J. A. and Jarvis, J. F. 1979, *Astrophysical Journal (Letters)*, **230**, L153.
Wall, J. V. 1983, in *The Origin and Evolution of Galaxies*, eds. B. J. T. Jones and J. E. Jones, p. 295. (Dordrecht: Reidel).
Weedman, D. W. 1980, *Astrophysical Journal*, **237**, 326.
Weedman, D. W. 1985, *Astrophysical Journal Supplement*, **57**, 523.
Wills, D. and Lynds, R. 1978, *Astrophysical Journal Supplement*, **36**, 317.
Windhorst, R. A., Kron, R. G. and Koo, D. C. 1984, *Astronomy and Astrophysics Supplement*, **58**, 39.
Windhorst, R. A., Miley, G. K., Owen, F. N., Kron, R. G. and Koo, D. C. 1985, *Astrophysical Journal*, **289**, 494.
Yee, H. K. C. and Green, R. F. 1984, *Astrophysical Journal*, **280**, 79.
Zuiderwijk, E. J. and de Ruiter, H. R. 1983, *Monthly Notices Royal Astronomical Society*, **204**, 675.

Chapter 3

Aaronson, M., Mould, J., Huchra, J., Sullivan, W. T. III, Schommer, R. A. and Bothun, G. D. 1980, *Astrophysical Journal*, **239**, 12.
Burbidge, G. 1979, *Nature*, **282**, 451.
Carney, B. W. 1980, *Astrophysical Journal Supplement*, **42**, 481.
Cocke, W. J. and Tifft, W. G. 1983, *Astrophysical Journal*, **268**, 56.
Davis, M., Huchra, J., Latham, D. W. and Tonry, J. 1982, *Astrophysical Journal*, **253**, 423.
Davis, M. and Peebles, P. J. E. 1982, *Astrophysical Journal*, **267**, 465.
de Vaucouleurs, G. 1979, *Astrophysical Journal*, **227**, 380.
de Vaucouleurs, G. and Bollinger, G. 1979, *Astrophysical Journal*, **233**, 433.
de Vaucouleurs, G. and Peters, W. L. 1981, *Astrophysical Journal*, **248**, 395.
Faber, S. M. and Gallagher, J. S. 1979, *Annual Reviews of Astronomy and Astrophysics*, **17**, 135.
Gorenstein, M. V. and Smoot, G. F. 1981, *Astrophysical Journal*, **244**, 361.
Guth, A. H. 1981, *Physical Review D*, **23**, 347.
Hayes, D. S. and Latham, D. W. 1975, *Astrophysical Journal*, **197**, 593.
Hoessel, J. G., Gunn, J. E. and Thuan, T. X. 1980, *Astrophysical Journal*, **241**, 486.
Hoyle, F. 1949, *Monthly Notices Royal Astronomical Society*, **109**, 365.
Hubble, E. P. 1929, *Proceedings National Academy of Sciences (US)*, **15**, 168.
Hubble, E. P. and Tolman, R. 1935, *Astrophysical Journal*, **82**, 302.
Kristian, J., Sandage, A. and Westphal, J. A. 1978, *Astrophysical Journal*, **221**, 383.
Peebles, P. J. E. 1971, *Physical Cosmology*. (Princeton University Press).
Rood, H. J., Page, T. L., Kintner, E. C. and King, I. R. 1972, *Astrophysical Journal*, **175**, 627.
Rowan-Robinson, M. 1977, *Cosmology*. (Oxford University Press).
Sandage, A. 1961, *Astrophysical Journal*, **134**, 916.
Sandage, A. 1975, in *Galaxies and the Universe*, eds. A. Sandage, M. Sandage and J. Kristian, pp. 761–83. (University of Chicago Press).
Sandage, A. 1982, *Astrophysical Journal*, **252**, 553.
Sandage, A. and Tammann, G. 1975*a*, *Astrophysical Journal*, **196**, 313.
Sandage, A. and Tammann, G. 1975*b*, *Astrophysical Journal*, **197**, 265.
Sandage, A. and Tammann, G. 1982, *Astrophysical Journal*, **256**, 339.
Schmidt, M. and Green, R. F. 1983, *Astrophysical Journal*, **269**, 352.
Segal, I. E. 1980, *Proceedings National Academy of Sciences (US)*, **77**, 10.
Tully, R. B. 1982, *Astrophysical Journal*, **257**, 389.
Weinberg, S. 1972, *Gravitation and Cosmology*. (New York: Wiley and Sons).

Chapter 4

Atwood, B., Baldwin, J. A. and Carswell, R. F. 1985, *Astrophysical Journal*, **292**, 58.
Barnothy, J. M. and Barnothy, M. F. 1972, *Astrophysical Journal*, **174**, 477.
Black, J. H. 1981, *Monthly Notices Royal Astronomical Society*, **197**, 553.
Boroson, T. A., Oke, J. B. and Green, R. F. 1982, *Astrophysical Journal*, **263**, 32.
Canizares, C. R. 1981, *Nature*, **291**, 620.
Canizares, C. R. 1982, *Astrophysical Journal*, **263**, 508.
Coleman, G. D., Wu, C.-C. and Weedman, D. W. 1980, *Astrophysical Journal Supplement*, **43**, 393.
de Vaucouleurs, G. 1959, *Handbuch der Physik*, **53**, 311.
Einstein, A. 1936, *Science*, **84**, 506.
Foltz, C. B., Weymann, R. J., Röser, H.-J. and Chaffee, F. H. Jr 1984, *Astrophysical Journal (Letters)*, **281**, L1.
Gehren, T., Fried, J., Wehinger, P. A. and Wyckoff, S. 1984, *Astrophysical Journal*, **278**, 11.

Gorenstein, M. V. *et al.* 1984, *Astrophysical Journal*, **287**, 538.
Heckman, T. M., Bothun, G. D., Balick, B. and Smith, E. P. 1984, *Astronomical Journal*, **89**, 958.
Huchra, J., Gorenstein, M., Kent, S., Shapiro, I., Smith, G., Horine, E. and Perley, R. 1985, *Astronomical Journal*, **90**, 691.
Hutchings, J. B., Crampton, D., Campbell, B., Duncan, D. and Glendenning, B. 1984, *Astrophysical Journal Supplement*, **55**, 319.
Kent, S. M. 1984, *Astrophysical Journal Supplement*, **56**, 105.
Kriss, G. A. and Canizares, C. R. 1982, *Astrophysical Journal*, **261**, 51.
Malkan, M. A., Margon, B. and Chanan, G. A. 1984, *Astrophysical Journal*, **280**, 66.
Petrosian, V. 1976, *Astrophysical Journal (Letters)*, **209**, L1.
Press, W. H. and Gunn, J. E. 1973, *Astrophysical Journal*, **185**, 397.
Refsdal, S. 1964, *Monthly Notices Royal Astronomical Society*, **128**, 307.
Sargent, W. L. W., Young, P. J., Boksenberg, A. and Tytler, D. 1980, *Astrophysical Journal Supplement*, **42**, 41.
Sargent, W. L. W., Young, P. and Schneider, D. P. 1982, *Astrophysical Journal*, **256**, 374.
Savage, B. D. and Jeske, N. A. 1981, *Astrophysical Journal*, **244**, 768.
Tinsley, B. M. 1976, *Astrophysical Journal (Letters)*, **209**, L7.
Turner, E. L., Ostriker, J. P., and Gott, J. R. III 1984, *Astrophysical Journal*, **284**, 1.
Turnshek, D. A. 1984, *Astrophysical Journal*, **280**, 51.
Tyson, J. A. 1983, *Astrophysical Journal (Letters)*, **272**, L41.
Tyson, J. A. and Jarvis, J. F. 1979, *Astrophysical Journal (Letters)*, **230**, L153.
Walsh, D., Carswell, R. F. and Weymann, R. J. 1979, *Nature*, **279**, 381.
Weedman, D. W. 1976, *Quarterly Journal Royal Astronomical Society*, **17**, 227.
Weedman, D. W. and Huenemoerder, D. P. 1985, *Astrophysical Journal*, **291**, 72.
Weymann, R. J., Williams, R. E., Peterson, B. M. and Turnshek, D. A. 1979, *Astrophysical Journal*, **234**, 33.
Wolfe, A. M., Turnshek, D. A., Smith, H. E. and Cohen, R. D. 1986, *Astrophysical Journal Supplement* (in press).
Yee, H. K. C. 1980, *Astrophysical Journal*, **241**, 894.
Yee, H. K. C. 1983, *Astrophysical Journal*, **272**, 473.
Young, P., Gunn, J. E., Kristian, J., Oke, J. B. and Westphal, J. A. 1980, *Astrophysical Journal*, **241**, 507.

Chapter 5

Adams, T. F. 1977, *Astrophysical Journal Supplement*, **33**, 19.
Fairall, A. P. 1984, *Monthly Notices Royal Astronomical Society*, **210**, 69.
Felten, J. E. 1977, *Astronomical Journal*, **82**, 861.
Filippenko, A. V. and Sargent, W. L. W. 1985, *Astrophysical Journal Supplement*, **57**, 503.
Fu-Zhen, C., Danese, L., De Zotti, G. and Franceschini, A. 1985, *Monthly Notices Royal Astronomical Society*, **212**, 857.
Gioia, I. M., Maccacaro, T., Schild, R. E., Stocke, J. T., Liebert, J. W., Danziger, I. J., Kunth, D. and Lub, J. 1984, *Astrophysical Journal*, **283**, 495.
Huchra, J. P. and Sargent, W. L. W. 1973, *Astrophysical Journal*, **186**, 433.
Huchra, J. P., Wyatt, W. F. and Davis, M. 1982, *Astronomical Journal*, **87**, 1628.
Keel, W. C. 1983, *Astrophysical Journal Supplement*, **52**, 229.
Kirshner, R. P., Oemler, A. Jr and Schechter, P. L. 1979, *Astronomical Journal*, **84**, 951.
Kirshner, R. P., Oemler, A. Jr, Schechter, P. L. and Shectman, S. A. 1981, *Astrophysical Journal (Letters)*, **248**, L57.
Kriss, G. A. and Canizares, C. R. 1982, *Astrophysical Journal*, **261**, 51.

Lonsdale, C. J., Helou, G., Good, J. C. and Rice, W. 1985, *Cataloged Galaxies and Quasars Observed in the IRAS Survey* (Pasadena: Jet Propulsion Laboratory).

Low, F. J. *et al.* 1984, *Astrophysical Journal (Letters)*, **278**, L19.

Maccacaro, T., Gioia, I. M. and Stocke, J. T. 1984, *Astrophysical Journal*, **283**, 486.

Markarian, B. E. 1967, *Astrofizika*, **3**, 55.

Markarian, B. E., Lipovetsky, V. A. and Stepanian, J. A. 1981, *Astrofizika*, **17**, 619.

Osterbrock, D. E. and Dahari, O. 1983, *Astrophysical Journal*, **273**, 478.

Piccinotti, G., Mushotzky, R. F., Boldt, E. A., Holt, S. S., Marshall, F. E., Serlemitsos, P. J. and Shafer, R. A. 1982, *Astrophysical Journal*, **253**, 485.

Rieke, G. and Lebofsky, M. J. 1979, *Annual Review Astronomy and Astrophysics*, **17**, 477.

Sargent, W. L. W. 1970, *Astrophysical Journal*, **160**, 405.

Schmidt, M. and Green, R. F. 1983, *Astrophysical Journal*, **269**, 352.

Stauffer, J. R. 1982, *Astrophysical Journal Supplement*, **50**, 517.

Tananbaum, H. *et al.* 1979, *Astrophysical Journal (Letters)*, **234**, L9.

Tarenghi, M., Tifft, W. G., Chincarini, G., Rood, H. J. and Thompson, L. A. 1979, *Astrophysical Journal*, **234**, 793.

Trumpler, R. J. and Weaver, H. F. 1953, *Statistical Astronomy* (New York: Dover Publications).

Weedman, D. W. 1985, *Astrophysical Journal Supplement*, **57**, 523.

Yee, H. K. C. 1983, *Astrophysical Journal*, **272**, 437.

Zamorani, G. *et al.* 1981, *Astrophysical Journal*, **245**, 357.

Chapter 6

Avni, Y. and Tananbaum, H. 1982, *Astrophysical Journal (Letters)*, **262**, L17.

Begelman, M. C., Blandford, R. D. and Rees, M. J. 1984, *Reviews of Modern Physics*, **56**, 255.

Carswell, R. F. and Smith, M. G. 1978, *Monthly Notices Royal Astronomical Society*, **185**, 381.

Carter, B. 1979, in *Active Galactic Nuclei*, eds. C. Hazard and S. Mitton, p. 185. (Cambridge University Press).

Cavaliere, A., Giallongo, E., Messina, A. and Vagnetti, F. 1983, *Astrophysical Journal*, **269**, 57.

Dahari, O. 1985, *Astrophysical Journal Supplement*, **57**, 643.

Danese, L., De Zotti, G. and Franceschini, A. 1985, *Astronomy and Astrophysics*, **143**, 277.

Elvis, M., Wilkes, B. J. and Tananbaum, H. 1985, *Astrophysical Journal*, **292**, 357.

Felten, J. E. 1977, *Astronomical Journal*, **82**, 861.

Gaston, B. 1983, *Astrophysical Journal*, **272**, 411.

Giaconni, R. *et al.* 1979, *Astrophysical Journal (Letters)*, **234**, L1.

Hazard, C. and McMahon, R. 1985, *Nature*, **314**, 238.

Hazard, C., Morton, D. C., Terlevich, R. and McMahon, R. 1984, *Astrophysical Journal*, **282**, 33.

Hutchings, J. B., Crampton, D. and Campbell, B. 1984, *Astrophysical Journal*, **280**, 41.

Keel, W. C., Kennicutt, R. C., Hummel, E. and Van der Hulst, J. 1985, *Astronomical Journal*, **90**, 708.

Kirshner, R. P., Oemler, A. Jr and Schechter, P. L. 1979, *Astronomical Journal*, **84**, 951.

Kriss, G. A. and Canizares, C. R. 1985, *Astrophysical Journal*, **297**, 177.

Lewis, D. W., MacAlpine, G. M. and Weedman, D. W. 1979, *Astrophysical Journal*, **233**, 787.

Maccacaro, T., Gioia, I. and Stocke, J. T. 1984, *Astrophysical Journal*, **283**, 486.

Marshall, F. E., Boldt, E. A., Holt, S. S., Miller, R. B., Mushotzky, R. F., Rose, L. A., Rothschild, R. E. and Serlemitsos, P. J. 1980, *Astrophysical Journal*, **235**, 4.

Marshall, H. L., Avni, Y., Tananbaum, H. and Zamorani, G. 1983a, *Astrophysical Journal*, **269**, 35.

Marshall, H. L., Tananbaum, H., Zamorani, G., Huchra, J. P., Braccesi, A. and Zitelli, V. 1983*b*, *Astrophysical Journal*, **269**, 42.
Marshall, H. L., Avni, Y., Braccesi, A., Huchra, J. P., Tananbaum, H., Zamorani, G. and Zitelli, V. 1984, *Astrophysical Journal*, **283**, 50.
Mathez, G. 1978, *Astronomy and Astrophysics*, **68**, 71.
McCray, R. 1979, in *Active Galactic Nuclei*, eds. C. Hazard and S. Mitton, p. 227. (Cambridge University Press).
Osmer, P. S. 1982, *Astrophysical Journal*, **253**, 28.
Peacock, J. A. and Gull, S. F. 1981, *Monthly Notices Royal Astronomical Society*, **196**, 611.
Petrosian, V. 1973, *Astrophysical Journal*, **183**, 359.
Rees, M. 1977, *Quarterly Journal Royal Astronomical Society*, **18**, 429.
Roos, N. 1985, *Astrophysical Journal*, **294**, 486.
Sandage, A. 1961, *Astrophysical Journal*, **133**, 355.
Sargent, W. L. W., Young, P. J., Boksenberg, A., Shortridge, K., Lynds, C. R. and Hartwick, F. D. A. 1978, *Astrophysical Journal*, **221**, 731.
Schmidt, M. 1968, *Astrophysical Journal*, **151**, 393.
Schmidt, M. and Green, R. F. 1983, *Astrophysical Journal*, **269**, 352.
Schneider, D., Schmidt, M. and Gunn, J. 1983, *Bulletin American Astronomical Society*, **15**, 957.
Schwartz, D. A. 1979, in *Advances in Space Exploration, 3, X-ray Astronomy*, eds. W. A. Baitz and L. E. Peterson, p. 453. (Oxford: Pergamon).
Tinsley, B. 1977, *Astrophysical Journal*, **211**, 621.
Wall, J. V., Pearson, T. J. and Longair, M. S. 1981, *Monthly Notices Royal Astronomical Society*, **196**, 597.
Weedman, D. W. 1985, *Astrophysical Journal Supplement*, **57**, 523.
Wills, D. and Lynds, R. 1978, *Astrophysical Journal Supplement*, **36**, 317.
Young, P. J., Westphal, J. A., Kristian, J., Wilson, C. P. and Landauer, F. P. 1978, *Astrophysical Journal*, **221**, 721.

Chapter 7

Angel, J. R. P. and Stockman, H. S. 1980, *Annual Review Astronomy and Astrophysics*, **18**, 321.
Avni, Y. and Tananbaum, H. 1982, *Astrophysical Journal (Letters)*, **262**, L17.
Balzano, V. A. 1983, *Astrophysical Journal*, **268**, 602.
Blandford, R. D. and Rees, M. J. 1978, in *Pittsburgh Conference on BL Lac Objects*, ed. A. M. Wolfe, p. 328. (University of Pittsburgh).
Boldt, E. 1981, *Comments on Astrophysics*, **9**, 97.
Cohen, M. H. *et al.* 1983, *Astrophysical Journal*, **272**, 383.
Condon, J. J. 1984, *Astrophysical Journal*, **284**, 44.
Condon, J. J., O'Dell, S. L., Puschell, J. J. and Stein, W. A. 1981, *Astrophysical Journal*, **246**, 624.
Cruz-Gonzalez, I. and Huchra, J. P. 1984, *Astronomical Journal*, **89**, 441.
Cutri, R. M., Rieke, G. H. and Lebofsky, M. J. 1984, *Astrophysical Journal*, **287**, 566.
Davidson, K. and Netzer, H. 1979, *Reviews of Modern Physics*, **51**, 715.
De Zotti, G., Boldt, E. A., Cavaliere, A., Danese, L., Franceschini, A., Marshall, F. E., Swank, J. H. and Szymkowiak, A. E. 1982, *Astrophysical Journal*, **253**, 47.
Gehrz, R. D., Sramek, R. A. and Weedman, D. W. 1983, *Astrophysical Journal*, **267**, 551.
Gioia, I. M., Maccacaro, T., Schild, R. E., Stocke, J. T., Liebert, J. W., Danziger, I. J., Kunth, D. and Lub, J. 1984, *Astrophysical Journal*, **283**, 495.
Grandi, S. 1982, *Astrophysical Journal*, **255**, 25.

212   References

Halpern, J. P. 1985, *Astrophysical Journal*, **290**, 130.
Henriksen, M. J., Marshall, F. E. and Mushotzky, R. F. 1984, *Astrophysical Journal*, **284**, 491.
Huchra, J. P. 1977, *Astrophysical Journal*, **217**, 928.
Impey, C. D. and Brand, P. W. J. L. 1982, *Monthly Notices Royal Astronomical Society*, **201**, 849.
Jones, T. W., O'Dell, S. L. and Stein, W. A. 1974, *Astrophysical Journal*, **192**, 261.
Kriss, G. A. and Canizares, C. R. 1982, *Astrophysical Journal*, **261**, 51.
Kriss, G. A. and Canizares, C. R. 1985, *Astrophysical Journal*, **297**, 177.
Kurucz, R. L. 1979, *Astrophysical Journal Supplement*, **40**, 1.
Larson, R. B. 1974, *Monthly Notices Royal Astronomical Society*, **166**, 585.
Maccacaro, T., Gioia, I. M., Maccagni, D. and Stocke, J. T. 1984, *Astrophysical Journal (Letters)*, **284**, L23.
Malkan, M. A. 1983, *Astrophysical Journal*, **268**, 582.
Meurs, E. J. A. and Wilson, A. S. 1984, *Astronomy and Astrophysics*, **136**, 206.
Miley, G. K., Neugebauer, G. and Soifer, B. T. 1985, *Astrophysical Journal (Letters)*, **293**, L11.
Miller, J. S. 1981, *Publications Astronomical Society of the Pacific*, **93**, 681.
Mushotzky, R. F. 1982, *Astrophysical Journal*, **256**, 92.
Osterbrock, D. E. 1981, *Astrophysical Journal*, **246**, 696.
Penston, M. V., Fosbury, R. A. E., Boksenberg, A., Ward, M. J. and Wilson, A. S. 1984, *Monthly Notices Royal Astronomical Society*, **208**, 347.
Petre, R., Mushotzky, R. F., Krolik, J. H. and Holt, S. S. 1984, *Astrophysical Journal*, **280**, 499.
Puetter, R. C., Burbidge, E. M., Smith, H. E. and Stein, W. A. 1982, *Astrophysical Journal*, **257**, 487.
Reichert, G. A., Mushotzky, R. F., Petre, R. and Holt, S. S. 1985, *Astrophysical Journal*, **296**, 69.
Richstone, D. O. and Schmidt, M. 1980, *Astrophysical Journal*, **235**, 377.
Rieke, G. H., Cutri, R. M., Black, J. H., Kailey, W. F., McAlary, C. W., Lebofsky, M. J. and Elston, R. 1985, *Astrophysical Journal*, **290**, 116.
Rieke, G. H., Lebofsky, M. J. and Wisniewski, W. Z. 1982, *Astrophysical Journal*, **263**, 73.
Rudy, R. J. 1984, *Astrophysical Journal*, **284**, 33.
Rybicki, G. B. and Lightman, A. P. 1979. *Radiative Processes in Astrophysics*. (New York: Wiley and Sons).
Scheur, P. A. G. and Readhead, A. C. S. 1979, *Nature*, **277**, 182.
Schmidt, M. 1970, *Astrophysical Journal*, **162**, 371.
Shuder, J. M. 1981, *Astrophysical Journal*, **244**, 12.
Smith, M. G., Carswell, R. F., Whelan, J. A. J., Wilkes, B. J., Boksenberg, A., Clowes, R. G., Savage, A., Cannon, R. D. and Wall, J. V. 1981, *Monthly Notices Royal Astronomical Society*, **195**, 437.
Smith, M. G. and Wright, A. E. 1980, *Monthly Notices Royal Astronomical Society*, **191**, 871.
Soifer, B. T., Neugebauer, G., Oke, J. B., Matthews, K. and Lacy, J. H. 1983, *Astrophysical Journal*, **265**, 18.
Sramek, R. A. and Weedman, D. W. 1980, *Astrophysical Journal*, **238**, 435.
Stein, W. A. and Soifer, B. T. 1983, *Annual Review Astronomy and Astrophysics*, **21**, 177.
Stothers, R. 1972, *Astrophysical Journal*, **175**, 431.
Tananbaum, H. *et al.* 1979, *Astrophysical Journal (Letters)*, **234**, L9.
Tennant, A. F. and Mushotzky, R. F. 1983, *Astrophysical Journal*, **264**, 92.
Tucker, W. H. 1975, *Radiation Processes in Astrophysics*. (Cambridge, MA: MIT Press).
Tucker, W. H. 1983, *Astrophysical Journal*, **271**, 531.
Ulrich, M. H. *et al.* 1984, *Monthly Notices Royal Astronomical Society*, **206**, 221.
Ulvestad, J. S. 1982, *Astrophysical Journal*, **259**, 96.
Urry, C. M. and Shafer, R. A. 1984, *Astrophysical Journal*, **280**, 569.

Weedman, D. W., Feldman, F. R., Balzano, V. A., Ramsey, L. W., Sramek, R. A., and Wu, C.-C. 1981, *Astrophysical Journal*, **248**, 105.
Wills, D. and Lynds, R. 1978, *Astrophysical Journal Supplement*, **36**, 317.
Windhorst, R. A., Miley, G. K., Owen, F. N., Kron, R. G. and Koo, D. C. 1985, *Astrophysical Journal*, **289**, 494.
Worrall, D. M. and Marshall, F. E. 1984, *Astrophysical Journal*, **276**, 434.
Zamorani, G. *et al*. 1981, *Astrophysical Journal*, **245**, 357.

Chapter 8

Baldwin, J. A. 1979, in *Active Galactic Nuclei*, eds. C. Hazard and S. Mitton, p. 51. (Cambridge University Press).
Baldwin, J., Phillips, M. M. and Terlevich, R. 1981, *Publications Astronomical Society of the Pacific*, **93**, 5.
Canfield, R. C. and Puetter, R. C. 1981, *Astrophysical Journal*, **243**, 390.
Capriotti, E. R., Foltz, C. B. and Peterson, B. M. 1982, *Astrophysical Journal*, **261**, 35.
Cohen, R. D. 1983, *Astrophysical Journal*, **273**, 489.
Collin-Souffrin, S. 1978, *Physica Scripta*, **17**, 293.
Davidson, K. and Netzer, H. 1979, *Reviews of Modern Physics*, **51**, 715.
De Robertis, M. M. and Osterbrock, D. E. 1984, *Astrophysical Journal*, **286**, 171.
Feldman, F. R., Weedman, D. W., Balzano, V. A. and Ramsey, L. 1982, *Astrophysical Journal*, **256**, 427.
Ferland, G. J. and Mushotzky, R. F. 1982, *Astrophysical Journal*, **262**, 564.
Filippenko, A. V. 1985, *Astrophysical Journal*, **289**, 475.
Grandi, S. 1983, *Astrophysical Journal*, **268**, 591.
Halpern, J. P. and Filippenko, A. V. 1984, *Astrophysical Journal*, **285**, 475.
Heckman, T. M. 1980, *Astronomy and Astrophysics*, **87**, 152.
Khachikian, E. Ye. and Weedman, D. W. 1974, *Astrophysical Journal*, **192**, 581.
Kwan, J. and Krolik, J. H. 1981, *Astrophysical Journal*, **250**, 478.
Lacy, J. H., Soifer, B. T., Neugebauer, G., Matthews, K., Malkan, M., Becklin, E. E., Wu, C.-C., Boggess, A. and Gull, T. R. 1982, *Astrophysical Journal*, **256**, 75.
MacAlpine, G. M. 1981, *Astrophysical Journal*, **251**, 465.
Mathis, J. S. 1983, *Astrophysical Journal*, **267**, 119.
Mushotzky, R. and Ferland, G. J. 1984, *Astrophysical Journal*, **278**, 558.
Osterbrock, D. E. 1974, *Astrophysics of Gaseous Nebulae*. (San Francisco: Freeman).
Osterbrock, D. E. 1977, *Astrophysical Journal*, **215**, 733.
Osterbrock, D. E. 1981, *Astrophysical Journal*, **249**, 462.
Osterbrock, D. E. 1984, *Quarterly Journal Royal Astronomical Society*, **25**, 1.
Penston, M. V. and Perez, E. 1984, *Monthly Notices Royal Astronomical Society*, **211**, 33p.
Phillips, M. M., Charles, P. A. and Baldwin, J. A. 1983, *Astrophysical Journal*, **266**, 485.
Rudy, R. J. and Puetter, R. C. 1982, *Astrophysical Journal*, **263**, 43.
Savage, B. D. and Mathis, J. S. 1979, *Annual Review of Astronomy and Astrophysics*, **17**, 73.
Shuder, J. M. 1982, *Astrophysical Journal*, **259**, 48.
Slipher, V. M. 1918, *Lowell Observatory Bulletin*, **3**, 59.
Steiner, J. E. 1981, *Astrophysical Journal*, **250**, 469.
Ulrich, M. H. *et al*. 1984, *Monthly Notices Royal Astronomical Society*, **206**, 221.
Whittle, M. 1985, *Monthly Notices Royal Astronomical Society*, **213**, 33.
Wills, B. J., Netzer, H. and Wills, D. 1985, *Astrophysical Journal*, **288**, 94.
Wu, C.-C., Boggess, A. and Gull, T. R. 1983, *Astrophysical Journal*, **266**, 28.

Chapter 9

Bartel, N., Shapiro, I. I., Corey, B. E., Marcaide, J. M., Rogers, A. E. E., Whitney, A. R., Capallo, R. J., Graham, D. A., Romney, J. D. and Preston, R. A. 1982, *Astrophysical Journal*, **262**, 556.

Begelman, M. C. 1985, in *Astrophysics of Active Galaxies and Quasi-Stellar Objects*, ed. J. S. Miller, p. 411. (Mill Valley CA: University Science Books).

Begelman, M. C., Blandford, R. D. and Rees, M. J. 1984, *Reviews of Modern Physics*, **56**, 255.

Bridle, A. H. and Perley, R. A. 1984, *Annual Reviews Astronomy and Astrophysics*, **22**, 319.

De Robertis, M. M. and Osterbrock, D. E. 1984, *Astrophysical Journal*, **286**, 171.

Elvis, M. and Van Speybroeck, L. 1982, *Astrophysical Journal (Letters)*, **257**, L51.

Feldman, F. R., Weedman, D. W., Balzano, V. A. and Ramsey, L. W. 1982, *Astrophysical Journal*, **256**, 427.

Ferland, G. J. and Mushotzky, R. F. 1982, *Astrophysical Journal*, **262**, 564.

Heckman, T. M., Miley, G. K., van Breugel, W. J. and Butcher, H. R. 1981, *Astrophysical Journal*, **247**, 403.

Holt, S. S., Mushotzky, R. F., Becker, R. H., Boldt, E. A., Serlemitsos, P. J., Szymkowiak, A. E. and White, N. E. 1980, *Astrophysical Journal (Letters)*, **241**, L13.

Koski, A. 1978, *Astrophysical Journal*, **223**, 56.

Kriss, G. A. and Canizares, C. R. 1982, *Astrophysical Journal*, **261**, 51.

Kriss, G. A., Canizares, C. R. and Ricker, G. R. 1980, *Astrophysical Journal*, **242**, 492.

Lawrence, A. and Elvis, M. 1982, *Astrophysical Journal*, **256**, 410.

Meszaros, P. 1983, *Astrophysical Journal (Letters)*, **274**, L13.

Meurs, E. J. A. and Wilson, A. S. 1984, *Astronomy and Astrophysics*, **136**, 206.

Miller, J. S. and Antonucci, R. R. J. 1983, *Astrophysical Journal (Letters)*, **271**, L7.

Mushotzky, R. and Ferland, G. J. 1984, *Astrophysical Journal*, **278**, 558.

Osterbrock, D. E. 1984, *Quarterly Journal Royal Astronomical Society*, **25**, 1.

Peimbert, M. and Torres-Peimbert, S. 1981, *Astrophysical Journal*, **245**, 845.

Phinney, E. S. 1985, in *Astrophysics of Active Galaxies and Quasi-Stellar Objects*, ed. J. S. Miller, p. 453. (Mill Valley, CA.: University Science Books).

Rees, M. 1984, *Annual Review Astronomy and Astrophysics*, **22**, 471.

Reichert, G. A., Mushotzky, R. F., Petre, R. and Holt, S. S. 1985, *Astrophysical Journal*, **296**, 69.

Rudy, R. J. 1984, *Astrophysical Journal*, **284**, 33.

Schmidt, G. D. and Miller, J. S. 1985, *Astrophysical Journal*, **290**, 517.

Serabyn, E. and Lacy, J. H. 1985, *Astrophysical Journal*, **293**, 445.

Shapiro, S. L. and Teukolsky, S. A. 1985, *Astrophysical Journal (Letters)*, **292**, L41.

Shuder, J. M. 1981, *Astrophysical Journal*, **244**, 12.

Shuder, J. M. 1982, *Astrophysical Journal*, **259**, 48.

Slipher, V. M. 1918, *Lowell Observatory Bulletin*, **3**, 59.

Telesco, C. M., Becklin, E. E., Wynn-Williams, C. G. and Harper, D. A. 1984, *Astrophysical Journal*, **282**, 427.

Tennant, A. F. and Mushotzky, R. F. 1983, *Astrophysical Journal*, **264**, 92.

Terlevich, R. and Melnick, J. 1985, *Monthly Notices Royal Astronomical Society*, **213**, 841.

Turnshek, D. A. 1984, *Astrophysical Journal*, **280**, 51.

Ulrich, M. H. *et al.* 1984, *Monthly Notices Royal Astronomical Society*, **206**, 221.

Veron, P., Lindblad, P. O., Zuiderwijk, E. J., Veron, M. P. and Adam, G. 1980, *Astronomy and Astrophysics*, **87**, 245.

Weymann, R. J., Scott, J. S., Schiano, A. V. R. and Christiansen, W. A. 1982, *Astrophysical Journal*, **262**, 497.

Wilson, A. S. and Ulvestad, J. S. 1982, *Astrophysical Journal*, **263**, 576.

Yee, H. K. C. 1980, *Astrophysical Journal*, **241**, 894.

INDEX

accretion disk, 134, 141, 151, 155, 194–8
active galactic nucleus, 71, 135–9
advanced X-ray astrophysics facility, 9, 41, 127, 155, 195
angular diameter, 65, 71–9

Balmer
  decrement, 170–2
  emission lines, 167–76
  self-absorption, 169
blazars, 22, 146, 149, 152, 195
blue bump, 151, 194
blue stellar objects, 20, 31, 39
brightness temperature, 158
broad absorption line quasar, 80, 201
broad-line region, 152, 176, 183–5, 195, 198–204

carbon lines, 178, 182
charge coupled devices, 9, 13–16, 24, 39, 76, 132
closed universe, 47, 52, 57
collisional excitation, 169, 179–81
correlation function, 35–7
cosmological age, 47, 54, 69
cosmological constant, 55
cosmological principle, 47, 56
covering factor, 199, 201
critical density
  nebulae, 181, 184
  universe, 56

deceleration parameter, 55–8
density evolution, 121–7
density parameter, 55–8
diffuse X-ray background, 9, 41, 128–30, 154
dust, 142, 149, 161, 171, 202, 204

Eddington limit, 134, 192
emission coefficient, 167, 176

equivalent width, 21, 62, 175

flat spectrum sources, 25
flat universe, 53, 57
flux
  definition, 3, 60–2, 72
  equations, 60–4
  limits in infrared, 17, 40
  limits in radio, 4, 25, 157
  limits in ultraviolet, 10
  limits in X-ray, 8, 9, 41
  optical units, 3
  radio units, 3, 159
  X-ray units, 3
forbidden lines, 178–82
Friedmann cosmologies, 45, 58, 116

galactic extinction, 23, 105
galaxies
  diameters, 74–9
  luminosity function, 111, 125
  near quasars, 31, 34
  nuclear luminosity, 106–8
  spectra, 73
  surface brightness, 74–6
  surface density, 33, 90
gravitational lens
  detection, 85, 91
  intensification, 88
  splitting, 86
$G(>R)$ function, 25, 157

High Energy Astrophysical Observatory 1, 7
High Energy Astrophysical Observatory 2, 8
high resolution imager, 8, 41
Hubble constant
  definition, 53
  value, 54–7, 69, 110
Hubble space telescope, 10, 15, 42, 91, 124, 137